BISON
BOOKS

D0061569

OTHER TITLES BY LOREN EISELEY
AVAILABLE IN BISON BOOKS EDITIONS

The Firmament of Time
The Invisible Pyramid
The Night Country

LOREN EISELEY

ALL THE STRANGE HOURS

THE EXCAVATION OF A LIFE

Introduction to the Bison Books Edition
by Kathleen A. Boardman

UNIVERSITY OF NEBRASKA PRESS • LINCOLN

⊗

First Bison Books printing: 2000
Most recent printing indicated by the last digit below:
10   9   8   7   6   5   4   3   2   1

Library of Congress Cataloging-in-Publication Data
Eiseley, Loren C., 1907–1977.
All the strange hours: the excavation of life / Loren Eiseley;
introduction to the Bison Books edition by Kathleen A.
Boardman.
p.   cm.
Originally published: New York: Scribner, 1975.
Includes bibliographical references (p.   ).
ISBN 0-8032-6741-X (pbk.: alk. paper)
1. Eiseley, Loren C., 1907–1977.   2. Anthropologists—United
States—Biography.   3. Authors, American—20th century—
Biography.   I. Title.
GN21.E45A3   2000
301'.092—dc21
[B]
99-085961

INTRODUCTION

Kathleen A. Boardman

All the Strange Hours is "the excavation of a life." Imagine for a moment archaeologist Loren Eiseley at an isolated site, digging up and dusting off artifacts from the layers of his life while a nearby group of people watches and comments. One of them, Eiseley's biographer, expresses some reservations about what Eiseley has done with the details of his life: "Reading *All the Strange Hours* is akin to viewing a surrealist painting," he whispers to the others. "Rarely are the portraits strewn across Eiseley's inner landscape placed in chronological order. Some are slightly askew, while others . . . are hung upside down, standing truth on its head."[1] Several of Eiseley's old friends are perplexed and a little hurt by Eiseley's life story. One believes his actions have been misrepresented. Another wonders aloud why Eiseley never mentioned him. "It's as if I didn't exist," he says.[2] A literary critic shakes her head as she declares, "Eiseley's 'autobiography' . . . is hardly an autobiography, but a fragmented collection of reflections on a life in science and at the boundaries of science."[3] Another onlooker agrees that Eiseley's life-excavation is "a very queer affair, but . . . no more paradoxical and no more fey than the personality behind it."[4] As the bystanders speak, the sky darkens and an icy wind begins to blow from the High Plains; they huddle closer together and turn up their collars against the cold as the furry shapes of animals begin to appear at the edges of hedgerows and along the horizon. Finally Eiseley looks up from his digging and smiles in the fading light. "I appear to know nothing of what I truly am," he says cryptically. "I feel impelled to deny everything and hide what is left."

The University of Pennsylvania's first Benjamin Franklin Professor did not write an autobiography in the Franklin style. It's true that *All the Strange Hours*, like Franklin's autobiography, is written by an old man, viewing his life from a vantage point near its end. Eiseley often mentions in it that his life is drawing to a close. Many old friends have died, and the people he remembers as young have aged in the decades since he last saw them. But, unlike Franklin, Eiseley does not provide an upbeat, orderly account of his path to success, designed to serve as a model for others. He dispenses with this traditional approach in his opening chapter, an account of a failed keynote speech: distracted by a photographer during his speech, his mind flashes back to his youth while a rat upstages him by cavorting in the spotlight. The rat, Eiseley tells us later, is the trickster, at the scene to humble the proud man. But Eiseley is not just being modest: the flashback shows him as a young Depression-era hobo and confirms his lifelong identification with the underdog and the fugitive, with the ones who do not have success stories to relate. Constantly aware of his flaws, illnesses, anxieties, missed chances, and failures, and often weary of his public accomplishments, Eiseley refuses to represent himself as a conventionally successful man. Yet he does still offer himself as an Everyman— a prototype of the fugitive, the wanderer, the stranger, and the lifetime seeker whose mind ranges constantly across miles and millennia.

All the Strange Hours is an intellectual autobiography. Eiseley's primary interest is not in the facts and official details of his life.[5] "I have played a certain visible role in life," Eiseley tells us, "but that my thoughts have often been elsewhere is quite apparent." He is fascinated with the mind: the way it moves back and forth through time, retrieves memories, synthesizes images, and composes fragments into meaning. More particularly, he tries to understand his own mind. The book's longest section, "Days of a Thinker," tells of Eiseley's student days and his later experiences as a writer, researcher, professor, physical anthropologist, and university administrator. It's not surprising that in this section he discusses the development of his key ideas. But "Days of a Drifter," which devotes many pages to his childhood and rail-riding days, is also intellectual autobiography. It accounts for the formation of his ideas on class, on identity, on human tendencies ("men beat men"), and on relationships between animals and humans. "Days of a Doubter" shows him as an old man; here he examines, discards, and reaffirms the ideas that have been important in his life. So, while Eiseley does not admit us very far into his private life or dwell on his public life and honors, he lets us into his mind and instructs us in its processes.

Although roughly chronological, *All the Strange Hours* is not a linear, comprehensive account. It is assembled out of loosely linked essays. Occasionally Eiseley provides explicit transitions by referring to events he has already narrated. Sometimes recurring images connect the chapters: a gambler with dice, prison escapees in the snow, a boy with little golden crosses. Finally, this memoir circles back on itself as it reaches, in Eiseley's words, "the end of which I spoke in the beginning." Eiseley himself calls attention to the gaps and repetitions in his story. From the beginning, he lets us know what to expect: strange hours, flashbacks and weird juxtapositions, potsherds from an excavation, mirror shards from his mother's dressing table, fragments of memory. "Everything in the mind is in rat's country," he writes. Memories don't die, but they are carried back and forth by neurons ("mental pack rats"), and we can retrieve them only in bits and pieces. That's why, Eiseley tells us sadly, nothing can ever "be again as it was." He invites us to see the brain as an artist's loft: we can move the pictures around, but we can't destroy or add to them. Nor can we decide what will enter our minds. Eiseley's metaphor of the mental garret also helps us glimpse his work as a writer: "One has just so many pictures in one's head which, after one has stared at them long enough, make a story or an essay. Beyond that one is helpless."

Perhaps the mind is unable to add new pictures, just as the unhappy child is powerless to alter his experiences. But the mind is anything but helpless when it comes to arranging these pictures into patterns that make sense. The mind is the great Synthesizer, with a powerful impulse to connect and organize. Eiseley himself was well equipped with a strong visual memory and a creative imagination; as a scientist, he learned to extrapolate a larger dynamic process from fragmentary evidence. From his research on evolution, Darwin, and the history of science, he became fascinated with the occasions when new paradigms—ways of organizing facts or describing the natural world—displace old ones. At such times, "the kaleidoscope through which we peer at life shifts suddenly and everything is reordered." Eiseley writes about a number of radical if not historic alterations in his own perspective. He also writes from experience of the occupational hazard of the dedicated library researcher: as "trail leads onto trail," the researcher may become obsessed with collecting material, so that his brain no longer cares "to organize this precious knowledge or fix it into a pattern." Likewise, the essayist and autobiographer must remember to pause in their collecting of memories so that they can make sense of them.

Collector-interpreters, whether they are scientists or memoirists, also have to exercise caution and respect. Such people dig deep for artifacts or memories, but they protect themselves and respect others by knowing when to stay at the surface. Eiseley describes his experiences with a single-minded expedition leader who kept insisting that his archaeologists "dig deeper," in spite of the burial sites that they were disturbing. Looking back, Eiseley regrets not having objected more strongly: "Men should discover their past. I admit to this. It has been my profession. Only so can we learn our limitations and come in time to suffer life with compassion. Nevertheless, I now believe that there are occasions when the earth tells our story just as well, when the tomb should remain hidden."

He also applies this caution to the autobiographer who is tempted always to unearth more memories: "To tamper with the past, even one's own, is to bring at times that slipping, sliding, tenuous horror which revolves around all that is done, unalterable, and yet which abides unseen in the living mind." In Eiseley's case, too much digging and too much remembering could also bring him face to face with his own bleak heritage: "the mad Shepards," his mother's family. "Be careful," he tells himself whenever he is about to delve into his darker thoughts or reexamine his memories of his mother and aunt. A mind inherited from the Shepard clan might abruptly cease its synthesizing activities and turn on itself in a schizophrenic episode. Better not to go there.

We can see, then, some sources of Eiseley's caution as an autobiographer. It also seems reasonable that he would shape a memoir that reflected his ideas about the mind and his attitude toward time. Still, if one sees autobiography as no more than a selection of "strange hours," and if some of those hours are dangerous to contemplate, why write a memoir at all? Perhaps *All the Strange Hours* itself suggests an answer to this question.

"Stranger, tell me a story!" In 1936, a big sailor sitting next to Eiseley on a train suddenly turned to him with this request. Smiling, Eiseley asked the sailor, Tim Riley, to tell his own story instead. In "When the Trouble Comes," we read about Riley's life, but it's not clear that Eiseley ever told Riley the story he asked for. If we think for a moment about *All the Strange Hours* as Eiseley's belated response to Riley's request, we can see what he does with the roles of *stranger* and *storyteller.*

Introduction

The very word "stranger" reverberates with importance for Eiseley, as he tells us several times. First, "stranger" is a "westernism," and Eiseley identifies himself as a man of the West come east to the city. "Born in the central plains, compacted out of glacial dust and winter cold," this man tells us that he belongs on the high plains of the last ice age, when humans and animals were simply fellow creatures. His solitary strangeness and loneliness may have grown out of his personal history, but they are also his heritage as a western American. His ancestors kept rolling west, neglecting the importance of community ties: "Short though the white man's history may be in these western towns, it is sometimes terrifying by its very evanescence," Eiseley writes. "Americans made a mistake they have been paying for ever since. In response to the Homestead Act they have been strung out at nighttime into a vast solitude." Eiseley's preoccupation with this solitude and evanescence propels *All the Strange Hours*, as it does much of his other work.

Alienation, though, is not a peculiarly western condition; it's a human condition. Just as the Americans who went west set up living patterns that ensured loneliness, so humanity has estranged itself from other life on Earth. "Man is a strange creature," declares Eiseley, and this statement reverberates with exasperation and awe. Estranged from the rest of the world, and even from their own bodies, humans nevertheless have fascinating ways of thinking. To be human is to be a stranger; thus, Eiseley, as a stranger, can also claim to be "every man." And he does.

Although personal and historical circumstances have caused some of his alienation, Eiseley makes it clear that he is also a stranger by choice. Early experience with violence and class conflict left him "free of mobs and movements." His discomfort while he was a provost with the student protest movements of the sixties may have reminded him of his early decision to be free from all groups of like-minded people. He describes "The Most Perfect Day in the World" as a day of timeless companionship among rootless men who never ask each other who they are or where they are from. The traveler—especially the drifter—meets singular people who come from nowhere and never ask questions. By keeping their names and histories to themselves, Eiseley's fellow wanderers allow him to imagine that they are "always appearing from some other century, entering and exiting, as it were, at will." He can create a place, time, and story for each of them.

In Eiseley's work, both human and animal strangers sometimes arrive with messages but rarely stay to get acquainted. The peculiar messenger may be a person with an unusual perspective, a scientist "dancing outside the ring" with a totally new idea, a doomed stray dog that expresses the preciousness of life, or a cat that seems to say, "I . . . regret the borders between us." "Stranger" is thus an appropriate persona for Eiseley, and the strangers he meets in his travel and research make evocative subjects.

Many writers, especially in Eiseley's time, took on the roles of wanderer and outsider, even if, like Eiseley, they had been rooted in a place and a profession for much of their adult lives. Eiseley is unusual in being a stranger in time as well as place. "I am an anachronism," he writes. As an autobiographer in his seventies, he understandably feels at times like a relic of the past, uncomfortable and perhaps unwelcome among people who are "with it." Still, his sense of being out of the proper time goes beyond that superficial discomfort. He describes himself as obsessed with time, a time-driven child grown into a time-absorbed archaeologist. Anthropology appeals because it provides a chance to move easily between eras. Archaeology allows him, as a member of a field party, to "remain hidden" in a "timeless land." His studies of evolution and its history constitute additional efforts to deal with time, and Eiseley finally declares that "all the sciences are linked by one element, time." Thus, his memoir deals not only with the nostalgia common in autobiography but also with his professional efforts to confront the inevitable and terrifying changes of life.

An anachronism, of course, can be an object from the future rather than a relic of the past—and Eiseley occasionally layers time in his writing so that past, present, and future seem to coexist. Observing strangers through the professional eyes of a physical anthropologist, he finds that "one may occasionally look back upon a fragment of the past, as in the shape of a huge brow ridge, or even see the unknown features of our far distant progeny prematurely peeping into existence." In back lots and rail yards and city streets he occasionally has visions of the future: ragged men and women scavenging in the debris of civilization, just as they picked over mastodon bones during the ice ages. Drawing on his own experience among the homeless, he visualizes himself among these "troglodytes," saying that "we would be here . . . when the city had fallen." Passages like these suggest a certain bizarre permanence in the midst of change; they also show Eiseley hard at work, dealing with time in its various guises.

As an anachronism, a stranger in time, Eiseley notes the occasional "perfect day" when time stops (or seems to stop) and he and people around him are "out of time, secret, hidden," with the "momentary illusion that [they] have won the game of life." He asserts a temporary power over time by arranging "all the strange hours" in his autobiography in a sequence that bends chronology. He plays around with time while writing in a genre that must inevitably confront the passage of time. Nevertheless, he admits that his narrative "is faltering" as time reasserts its power: his readers, after all, expect a story to move along. Time is an arrow: "either you ride away inexorably upon its back, or, if you stop, it goes by you with someone else waving farewell whom you will never meet." Time, the dice-rolling gambler, is an unbeatable opponent, even for an anachronism.

A stranger in time and place, Eiseley tells us a story. While the role of stranger sets people apart, the act of storytelling brings them together. As an anthropologist, Eiseley knows that storytellers make an effort against both solitude and evanescence. They function to draw the community together—by gathering people for an entertainment, by saying "we," by speaking of common values, and by inspiring the group to evaluate their actions. Stories can memorialize people and events of the past, and perhaps the stories themselves are remembered or passed on, thus preserving the present for the future. Eiseley even suggests that his childhood might have been less fragmented if someone had been around to tell the stories of his own family and his own place. But instead, he says, "The names [of my ancestors] lie strewn in graveyards from New England to the broken sticks that rotted quickly on the Oregon trail. . . . How, among all these wanderers, should I have absorbed a code by which to live?" Rootless people, too, need a storyteller to draw them together—and who better than another member of their own community of wanderers?

Eiseley represents himself as a self-reliant western raconteur with a repertoire of stories that westerners like: dog stories, cat stories, rat stories, ghost stories, gambler stories, drifter stories. Many of them have a strangeness, a sense of the ridiculous and out-of-proportion that evokes the western tall tale. The story of Night Country, the talking cat, is a tall tale at one level, even though it broods upon the barriers between humans and animals. Although the tone of *All the Strange Hours* is serious, it's not out of the question that Eiseley is pulling our leg with some of his stories. The book also hints that Eiseley's life-storytelling is an oral as well as a written performance. Not only is his opening chapter a story about a speech gone wrong, but he occasionally addresses his audience directly, as in this passage: "I am, it is true, wandering out of time and place. This narrative is faltering. . . . Listen, or do not listen, it is all the same." His technique is not to coax his listeners but to turn away with a shrug, hoping that curiosity will impel us to follow him further into his story.

As a storyteller who is also a writer, Eiseley sometimes seems at odds with his audience, as if he believes we are criticizing him for an unconventional performance as a memoirist: "There was not time for everything. One must not be boring, but the essence is here. Here are the sounds my father, the actor, heard and spoke beautifully. Here is the beauty my deafened mother could see with her eyes alone and strive to paint. Perhaps you do not care for these things, my friends, but I care and I have come a weary distance. My anatomy lies bare. Read if you wish, or pass on." Such a passage sounds a bit pugnacious, even peevish. But Eiseley addresses us as friends and insists that he has told us about the important things in his life, the things he cares about. Like many memoirists, he believes he has laid himself bare, in his own fashion. He offers us the role of physical anthropologist: if we wish, we can study his bones as he studied the bones of others. In his disclaimer ("There was not time for everything") he cleverly reminds us of his preoccupation with time. As both writer and storyteller, he must have an audience. He must also have his own way.

While apparently relishing his persona as loner, stranger, and fugitive, Eiseley often expresses gratitude to the helpers who got him through rough spots, even saved his life. His thank-you to his wife, though brief, is touching. But he dismisses with annoyance all attempts to locate his *literary* helpers—any writers who influenced him or editors who helped shape his work: "If I say . . . that I have read Thoreau, then it has been Thoreau who has been my mentor. . . . Or it is Melville, Poe, anyone but me. If I mention a living writer whom I know, he is my inspiration, my fount of knowledge."

While assembling material for *The Immense Journey*, Eiseley tells us, he developed the idea of the "concealed essay." Free of audience concerns (since he thought he wouldn't have an audience anyway), he consciously pursued this form, "in which personal anecdote was allowed gently to bring under observation thoughts of a more purely scientific nature," without harming the scientific data. He persevered with the concealed essay, and response was divided: awards and praise from lay readers, criticism from some colleagues in his academic discipline. These experiences may help explain both Eiseley's ambivalence about his audience and his insistence on receiving the credit for his own development as a writer.

Although in other roles (as a son, drifter, research assistant, provost) Eiseley may have lacked power or failed to make a difference, he insists on the powers of the writer. The writer synthesizes fragments of observations and memories. The writer shines a spotlight on events that might go unnoticed: "This day I have recounted is gone from the minds of everyone," Eiseley writes about his account of a dog that was killed to serve as a lab demonstration for inattentive students. The writer records and memorializes. Describing a watchman who died years ago, Eiseley says, "I doubt if anyone else remembers Willy now, but I do . . . I am Willy's last recorder." The passages sound nostalgic, but they also assert a writer's strength in the battle against time and loneliness. Through his writing, this descendent of nameless and forgotten wanderers chooses which of his friends and enemies will be named and which will not. In *All the Strange Hours*, his own memorial, Eiseley revels in his ability to record the otherwise unrecorded.

But even the writer's powers don't last forever. In his final chapters, Eiseley meditates on disintegration and on his own approaching end. The mental pack rats that have carried and assembled his memories now seem to be dispersing him. "I who carried the memory was also dissolving," he tells us. "Trade rats were exchanging me piece by piece." As he notes ruefully in the book's first chapter, his story might be entered into some folklorist's collection as "the rat cycle." Although well aware that his writing will outlast him, he relinquishes his writer's roles to the pack rats, "who keep the record" in their middens. Eiseley's reflections on death seem inseparable from his thoughts about life. Life is a gamble against time, he writes, a throw of the dice. Life is a mystery, for after a lifetime of science, he has found "nothing to explain the necessity of life, nothing to explain the hunger of the elements to become life." Life is a dynamic process, like a stream that rushes out of the mountains, depositing stones and boulders in an alluvial fan: "We are not to be found among the stones [of memories], we have been the stream. And it is the stream, not the colliding boulders, that make up a life. . . . What eventually lies on the outwash fan is memory, and it is from memory that we hesitantly try to reconstruct the nature of each individual torrent." In *All the Strange Hours*, Eiseley leaves us vivid descriptions of carefully selected boulders in the outwash fan of his life. Lucky for us, he also provides a glimpse of the stream itself.

NOTES

1. Gale E. Christianson, *Fox at the Wood's Edge: A Biography of Loren Eiseley* (New York: Holt, 1990), 426.

2. Quoted in Peter Heidtmann, *Loren Eiseley: A Modern Ishmael* (Hamden CT: Archon, 1991), 15.

3. Mary Ellen Pitts, *Toward a Dialogue of Understanding: Loren Eiseley and the Critique of Science* (Bethlehem PA: Lehigh University Press, 1995), 33.

4. James Olney, "All the Strange Hours" (review), *New Republic*, 1 November 1975, 31–32.

5. Readers looking for these details and for the memories of the people who knew him should turn to Gale Christianson's excellent biography.

Dedicated to Charles Frederick Eiseley
cavalryman in the Grand Army of the Republic
member of the first legislature in
Nebraska Territory,
without whom I would not be here,
and
to William Buchanan Price
born in the ruins of the Confederacy
finally to lie in state
in the Capitol Rotunda of Nebraska,
without whose help
my life would have been different
beyond imagining

I' the color the tale takes, there's change perhaps;
'Tis natural, since the sky is different,
Eclipse in the air now, still the outline stays.

—Robert Browning

CONTENTS

PART ONE

DAYS OF
A DRIFTER

There is nothing worse for mortal men than wandering.
—*The Odyssey*

CHAPTER 1

The Rat That Danced

WHEN my aunt died I found among her effects a beautiful silver-backed Victorian hand mirror. It had been one of a twin pair my maternal grandfather had given to his girls. The last time I had seen my mother's mirror it had been scarred by petulant violence and the handle had been snapped off. It had marked the difference between the two girls—their care of things, perhaps their lives. I had looked into the mirror as a child, admiring the scrollwork on the silver. Mostly things like that did not exist in our house. Finally it disappeared. The face of a child vanished with it, my own face. Without the mirror I was unaware when it departed.

Make no mistake. Everything in the mind is in rat's country. It doesn't die. They are merely carried, these disparate memories, back and forth in the desert of a billion neurons, set down, picked up, and dropped again by mental pack rats. Nothing perishes, it is merely lost till a surgeon's electrode starts the music of an old player piano whose scrolls are dust. Or you yourself do it, tossing in the restless nights, or even in the day on a strange street when a hurdy-gurdy plays. Nothing is lost, but it can never be again as it was. You will only find the bits and cry out because they were yourself. Nothing can begin again and go right, but still it is you, your mind, picking endlessly over the splintered glass of a mirror dropped and broken long ago. That is all time is at the end when you are old—a splintered glass. I should never have gone to that place, never accepted the engagement, never have spoken under the lights of their brand-new auditorium.

Nineteen seventy-four was the first year I ever thought of myself as old. There was not so much ahead anymore and a lot left behind. I was remembering and not sleeping so well and up front in the future the view was getting foreshortened. Maybe I was reading the obituaries of old friends in the professional journals too often. I used never to pay attention to things like that. They were for somebody else. Now Teddy McCown was gone, Gene Vanderpool, Smitty Smith, Claude Hibbard, and those others. A lot of time had gone with them. I stood on the street of that damned Texas town and the sign on the door before me read:

> Anybody who objects to the sight of
> Nude People Making Love
> doesn't belong in here.
> Anything San Francisco can do
> we can do better.

Couldn't they ever stop playing chicken in Texas? Well, I could. I walked by the gun shop, and by the cop on the corner, and went into the finest hotel in town, to an air-conditioned room.

I drew the shades and lay down carefully. The glare on that harsh landscape was frightful. My heart felt tired and the whole convention was turning sour before it began. I had flown two thousand miles to make a speech, not my kind of speech, but the kind they had insisted they wanted. Next year, the officers of the association would be telling another speaker that last year's speech hadn't gone over so well with the delegates.

"Now, you're the man who—"

THE RAT THAT DANCED

Yes, I had heard it, looking into the mouthpiece with the qualms of experience, but I needed the money. They shouldn't have needled me like that. It was bad psychology. I felt like an old fighter who suddenly knows he has seen too many arenas. They hadn't arranged it the way I wanted. I lay back in the darkened room and listened to the youngsters splashing in the pool outside my window. The airport had been intolerable. So was the sun. Likewise the wealth. To hell with it, don't think, just do it when the time comes. Always you didn't sleep but you did it. Do it again, that was all. Do it after hot coffee in the morning, while people talked and you did not listen. I swallowed two sleeping pills. The morning would come fast enough.

The moment came late, I noted mentally from the wings. A bored audience already settling back from earlier events. No help for it and maybe a tired speaker in spite of his black coffee.

The stage was vast and made for greater things than myself. It was an enormous cavern with only the small yellow glow of a light on the podium. What was wrong? I was beginning to tread as on a rope over an infinity of time. I made a wry joke, more to myself than the audience. I started my speech. I was talking about time—time as it had existed in five great civilizations. I spoke of the fear of Columbus's sailors approaching the New World and their horror of falling off the edge of the world.

But it was not the world we fell from, I went on, momentarily identifying with the past. We toppled instead off the end of time. We, all of us, western man clinging in his little enclaves to the eastern seaboard, had re-entered the stone age. We had starved helplessly in our first winters; Indians had fed us. Generation by generation we had had to relearn the arts of a vanished era. In order to survive we had had to master what our paleolithic forebears had taken for granted. The farther we pressed into the forest the more rank, prestige, and fine garments would dissolve into rags and buckskin. We would be reduced to elemental men.

At this moment the press chose to make its appearance. From

out of the dark cavern a stealthy figure crept along the edge of my vision. Suddenly from his hands emerged a light so blinding I almost staggered. Under its impact I was falling backward across decades. I clung to the podium as to a door being ruthlessly forced open against my will. Where was I? Into that calm discourse there cut words emphasized by a pistol shot.

"Put up your hands," echoed a far-off voice.

"Face the wall."

"Search him."

There followed the swift professional patter of hands over my body.

"What's this?"

"A key."

"Ah, to what? Answer."

"My home. I had one once."

"So you say. All right, you can come along now. Wait till morning with the rest of your sort. Quit eyeing that express, you've ridden far enough. She's going without you."

The light on the stage faded though the afterimage stayed. The audience seemed insubstantial, vaporous. I tried again.

"All the sciences," I said desperately into the microphone, "are linked by one element, time. It pervades them all. When western Europe believed the world was six thousand years old the Maya had already set stelae to mark the passage of eras. The priesthood calculated in millions of years of vertical time. As someone put it,

> behind nothing
> before nothing
> worship it the zero . . ."

I broke off. The light had come again, this time a red glow at the stage edge. I was back by a switchlight in Sacramento opening a letter. I already knew what was in the letter. I had carried it in my pocket all the way from San Francisco but I was moving east again in answer to the summons that had come. "Your

father is dying, come home." All over America men were drifting like sargasso weed in a vast dead sea of ruined industry.

I picked some apples in an orchard near the grade where the
freights began their climb toward the high Sierras. Or was I
still on this Texas podium following time across America? I
was talking but the audience was receding, wavering again.

I was tied by a cord at my right wrist to a freight top laboring
eastward under the rising sun. I glanced along a mile of cars on
which scores of ragged men were rousing in the dawn.

"Listen, fellows," pleaded a brakeman coming down the line
of cars. "Just keep down, can't you? We've got to be able to see
the signals from the caboose to the engine." It seemed like half
of America was on that train.

I stood, for a while, like those others, stiff from a night on the
roofs. Mile after desolate mile, the Nevada desert glided by.
There were no more orchards or the cold tunnels of the high
Sierras. I gnawed an apple to quench my thirst.

Somewhere I lost the freight and rode the blinds of a mail
train from night into another day. A brakeman must have seen
me in the dawn. I missed the first blind, a safer perch, and
swung aboard after the mail car had passed.

The train was hitting sixty when the brakeman unlocked and
burst out of the following car. He was drawing on gloves and
he meant business.

"Get the hell off. Jump, you bastard," he roared, shouldering
me and trying to push me out above the pounding wheels.

"Look," I said, "you want to kill me?"

For answer he struck me across the face and pushed. A thin
hot wire like that in an incandescent lamp began to flicker in
my brain.

This man was not a cop, he was a worker out of my father's
time. He was trying to kill me for no reason but the sheer pleasure of it. I raised my shoulders as he battered at my head, taking care with my footing lest I should fall.

He was making a mistake. I was in my late teens but I was

all bones and rawhide, and murder was licking like flame along the edges of my mind. There was no mercy in him, and now there was none in me. While he slugged at me I went into a boxer's crouch and clung with one hand to the rocking platform. "Kill him, kill him," blazed the red wire. "He's trying to kill you."

I inched a tentative step forward while my forearm and cheekbone took his blows. He was a fool. All it would take was a slight shift of footing and the use of my right hand now clinging to the uprights. He would be gone, as he had wanted me to go, under the wheels. Around the pulsing red wire sanity was going, but it was not quite gone.

We were traveling one way across an utter, inhospitable desert. There was no water. I had to stay by the water tanks. If he went under they would have me in hours. There was no mercy on that rocking, speeding mail train, none in my mind. I could have killed him. I could kill him now after all these years. It is thus one learns the depths of hatred. But some fragment of sanity remaining in my head was wary. Being of a bookish turn of mind, I had found on the road only timetables to study. "This is a short run. Be careful not to let him shove you through the stanchions," a whisper went through my mind. "He can't hurt you, he's out of wind and pounding himself out."

Over my guard I snarled at him, but already the train was slackening speed. He rushed inside as suddenly as he had descended upon me. I didn't wait. I knew the instant the train stopped, he would set the railroad police on me. I edged carefully between the uprights, hung a moment more from the ladder, then jumped. I rolled once and came up running. I didn't even stop to feel my face. I was over a fence, out of railroad property and vanishing.

Hours later among dozens of other outcasts I washed my bruises and my clothes in a hobo jungle at Provo, Utah. The stream made a merry little sound.

The wire, the hot, red murderous wire, still pulsed somewhere in my brain. My father was a mild-mannered man with a deep faith in the essential goodness of the working class.

"Who slugged you, kid?" remarked a grizzled, broken-nosed man with whom I was destined to travel onward.

"A brakie on an express," I told him. He listened, unsmiling, imperturbable.

"There'll be a freight along here in a few hours," he said. "Want to travel with me?"

"Sure." I stood up and tightened my belt.

"Safer that way," he remarked cryptically. Then he put his hat over his eyes and went to sleep.

"Careful, fella," someone else whispered to me as I nursed a battered cheekbone. "That guy's got a gun. He's dangerous."

"Well," I said, "I've got nothing, so he can't be using it on me."

That night we made it over the Rockies in the empty ice compartment of a refrigerator car. I awoke in the dawn with a two-hundred-pound slab of ice poised on tongs above my head and someone peering down. "Where are we?" I asked the good-natured icer as we climbed stiffly out. "Cheyenne," he answered grinning. "You're over the hump. It's all downhill now, but watch out for the yard bulls."

My companion grunted.

Days later we sat by a fire together. "I'm heading down into the harvest fields," said the grizzled man. I knew beforehand there was always a time of violence in the wheat. Stickups began when the migrant harvest workers started to move with the trains.

"I'm going on," I said. "I have to." He looked at me with the eyes of middle age, prison eyes, black, impenetrable, self-sufficient. I knew by then that if he had a gun it was not on him. It was hidden somewhere in a coal car that his kind had a way of finding again at division stops.

He tilted his hat, indifferently. Men parted this way on the road, but we had shared some hours and food together. It was also a time when radical talk, half real, half fantasy, was spun out beside hundreds of fires across America. We dug into a final sack of food and shared it.

"Face all right now?" he queried, though it was evident he was already far off somewhere and receding from me.

"Yes," I said and stopped. What was there to thank him for? Company? He wouldn't have understood.

The sack was emptied. He stood up in the firelight and cast it on the flames. The paper flared briefly, accentuating the hard contours of his face. "Remember this," he said suddenly, dispassionately, as though the voice originated over his shoulder. "Just get this straight. It's all there is and after a while you'll see it for yourself." He studied me again without expression. "The capitalists beat men into line. Okay? The communists beat men into line. Right again?"

"I reckon," I ventured, more to fill in the silence growing around us than because I understood.

He pointed gently at my swollen face. "Men beat men, that's all. That's all there is. Remember it, kid. Take care of yourself." He walked away up the dark diverging track.

That man, whose name I never knew, must be long dead. I know he would have died as he had lived, perhaps in his final moments staring silently upward at the cracked ceiling of a Chicago flophouse, or alone in some gun-lit moment of violence.

Years later when the bodies of men like him lay on dissecting tables before me, I steeled myself to look at their faces. I never found him. I am glad I never did, but if I had I would have claimed him for burial. I owed him that much for some intangible reason. He did not kill the illusions of youth, not right away. But he left all my life henceforward free of mobs and movements, free as only very wild things are both solitary and free. I owed him that.

> Before nothing
> behind nothing
> worship it the zero.

THE RAT THAT DANCED

A spatter of applause was sweeping my audience. I was still clutching the podium as I had clung to the flying train. Looking around I tried to determine what I had said. I felt as battered as on that night of fifty years ago.

"Did you see him, did you see him?" exclaimed someone from the audience, rushing forward.

"No, what?" I said, still returning from the glow of a long-vanished fire and drawing a fist down my face.

"The rat," explained the man, whose features now registered in my mind as a former and distinguished student.

"Nice to see you, Dick," I said weakly. "A rat, you say?" Others were grinning.

"The spotlight," Dick said. "It caught him right out of the dark. He tumbled, he ran, he played, he danced. It was fantastic, you should have seen it. Imagine, in a big municipal auditorium!"

"My God," I grumbled, still feeling my face. So that was the reason for the ironic applause. My speech, I knew, with an inner listening ear, except for the poem, had become increasingly incoherent in the white glare that blinded me. The applause was not for me. It was for the rat, the rat that danced.

Well, I thought, as I walked toward the entrance and the waiting reporters, you asked for it, and thanks, Dick, for telling me. Otherwise I would not have guessed. Wasn't it I who had once written that there was a trickster in every culture who humbles what are supposed to be our greatest moments? The trickster who reduces pride, Old Father Coyote who makes and unmakes the world in a long cycle of stories and, incidentally, gets his penis caught in a cleft pine for his pains.

I had summoned a rat who danced and walked out to incredulous stares and applause. The rat cycle. Some folklorist could have it now. Something for Texas. It fitted me for accepting. The rat who danced. Only one of my own students would have caught the irony. Good old Dick who knew the ways of the Havasupai. I laughed but the trickster always brings pain. Why had I assented indifferently to the photographer with his police-state equipment? Damn them all, they had called back the past. It was I, I who had been the rat that danced, the rat by reflex, the Pavlovian rat activated by a light. The tale I had intended of the great dead cities that might be restored as shrines of meditation had been lost in the incoherence of a split personality, the murderer who had not murdered but who carried a red wire glowing in his brain.

While I tossed sleepless, waiting for morning, the rat danced at the periphery of my vision, an afterimage that could not be exorcised—not till long, long afterward. The trickster who humbles pride. Every funeral has one. Only the old people from the silent cliff houses and the horse people of the plains had known and institutionalized him—the backward dancing man, the caricaturist of order. As for me, the rat went on dancing, dancing in my wearied brain as I rolled upon the pillow. *Behind nothing, before nothing, worship it the zero.*

No, I had worshipped enough. Toward morning I dreamed in the half light. There was a train, a train of long ago picking up speed at a crossing. On impulse I waved from the car roof at a girl in a red roadster stopped at the barrier. She waved back, but I knew I would never see her again. Everything was moving. It was autumn, the wind whirling the leaves. The train was picking up momentum on the grade. I waved till we rounded a curve. There was no going back. It was a brief, heartsick moment of revulsion. Time's arrow, I would later learn to call it in the lecture halls of universities. I wondered where she was going in that car. I wonder where she is now and if she remembers time itself rolling by, and the man with goggles clinging to the freight top. Time. Either you ride away inexorably upon its back, or, if you stop, it goes by you with someone else waving farewell whom you will never meet. I left the airport on the first morning flight to the East.

CHAPTER 2

The Life Machine

I T was the last day. I stood in a corner of the room and watched him die. For hours there had been no sign of consciousness. A nurse intervened. She shouted in his ear. "Your son Leo is here. Leo has come. Leo, Leo is here."
Leo was my half-brother, fourteen years older than I, the son of an earlier marriage. Leo's mother was dead. I was the child of a second marriage, long after, in Lincoln, Nebraska.

Slowly, to my boundless surprise, the dying man's eyes, indifferent to me for many hours, opened. There was an instant of recognition between the two of them, from which I was excluded. My father had come back an infinite distance for that meeting. It was wordless.

I walked out into the hall unnoticed. It was only just, I thought fleetingly without rancor. Leo was the son of my father's youth, of a first love who had perished in her springtime and of whom my father could never bring himself to speak. I was born when father was forty, of a marriage that had never been happy. I was loved, but I was also a changeling, an autumn child surrounded by falling leaves. My brother who had been summoned was the one true son, not I. For him my father had come the long way back, if only for a moment.

In the corridor I leaned against the wall, adjusting without heat, without passion, to what I had always known. There was an equality of years and understanding, of secrets shared, that I would never have the opportunity to enjoy. I was too young, too violent, and now he was going. In that hall, in that room, I would not weep before others. I had already learned that much. Perhaps they might have tried to call my father back a second time with my name. I did not request it. A man can do just so much, whether he is a father or a son. In a little while he was gone.

He left a single thumbed copy of Shakespeare inscribed with his name, which I still own. A few letters lingered in the possession of my aunt, with whom I was staying at the time. She was a kind woman, but she had a morbid dread of disease. A week or so after the funeral I came upon her incinerating in the yard the remaining letters. Perhaps it was just as well. As I stood beside her a charred sheet turned over. Written across it in the fine bold penmanship of my father was a sentence rapidly being obliterated in the flames. I saw it curl and crumble as I read.

"Remember, the boy is a genius, but moody." Whatever reservations the letter contained would never be read by me. There was only the little licking flame and the words curling up and going black. I could never reread them. Of course I was no genius, of course I was moody, as anyone young would have been under the restraints of that household. Nevertheless, the words were proudly comforting. My father had recognized me after all. In the limitations of that day and time he had always defended my right to read books. He had lived just long enough to think the books would lead me to someplace unknown. He had that kind of mind. Whatever else the letter had contained or in what context the remark had appeared was now irrelevant. He had known, however overstated, that I *was* a changeling, an oddity in the cradle of a belated second marriage.

By then he was a man worn with grief and labor, cast out on the industrial scrap heap in the day of the locusts. *"The capi-*

talists beat men into line," ran bitterly through my head. Father was, at the time of his death, a traveling hardware salesman spending solitary nights among the artificial palms of small-town hotels. What he thought, whether he dreamed alone of the brief years of his first marriage, I would never know.

The depression struck. Like thousands of other aging, un-aggressive salesmen he was expendable. His dismissal had been abrupt and brutal. He came home yellow and cadaverous. Already, though he did not speak of it, he must have known the truth: cancer. By the time I arrived he was lying on a couch unable to eat and the black exudate of blood was beginning to well up from his stomach.

He was removed to the hospital by a callous ex-Army surgeon who infrequently visited his bedside. He died by inches. I had had no experience of this and entertained a slightly misplaced faith in the humane nature of medical procedure.

Looking back I sometimes wonder, with a pang of conscience, if I should have accepted the consequences and shot him. Still, did that last glance at my far-traveled brother justify the waiting? I am too far removed in time to decide. At that moment only youth, the lack of a gun, and indecision were the real factors.

While still able to walk he had called me into the bathroom to consult me upon a strange growth already sprouting from his side. "Christ," I thought, as I groped to reassure him, "he's riddled with it, he's done for." I patted his shoulder. "Pa," I had said, "the hospital will fix all that. Just lie down and take it easy now." So he lay there while we waited and the black exudate ran from his lips.

"Get me a pickle, Loren," he had said. "I got a craving for something sour to suck on." "Yes, pa," I had said and had gone out and sent a wire to my brother. Then I had stepped into my uncle's garage alone and put my head against a beam and wept, furiously and briefly. It was, I think, the last time I ever wept, except long, long afterward about a woman.

The crumpling flame-embroidered words in the fire haunted me simply because they were the last message I would ever have from my father. But it was no time for books. Through a chance connection I picked up a job in a chicken hatchery.

The place was a potential firetrap. Straw for the hatched chicks and drums of kerosene for the incubator heaters were strewn along one side of the wooden building. I was supposed to be night security. Every hour on the hour I walked down the rows of huge incubators filled with sleeping life. I read the temperatures, turned trays filled with hundreds of eggs, checked the fuel in the lamps. Today I suppose it is all done electronically from a control room. In that time incubators the size of rooms had to be personally attended. When I had checked each one I went back to the office, set an alarm clock for the next hour, and dropped my head on my arms.

This was a routine I had devised for myself. Where I was supposed to sleep in the day, children played in the yard next door. This spasmodic sleep at night was all I was getting and it was horribly debilitating.

The giant incubators with their hidden, dormant life seemed menacing in the silent corridors. One had the feeling that something, someone not human, was walking down the aisle beyond, keeping pace with one's own steps, challenging even one's thoughts in the great humming hive.

Once I turned into a corridor just in time to see flames from a leaking heater creeping up the side of an incubator. In another moment the place would have been an inferno. If I had been one aisle away on my rounds I would never have survived to reach a door. I seized an extinguisher from a nearby beam and doused the fire, leaving only some blistered paneling on the incubator's surface. I half expected the creature in the next corridor to investigate as I stood there panting. I cleaned up everything and waited, listening. Nothing came. I inserted a new jug of kerosene, got the temperatures adjusted again and went back to set my clock in the office.

In the morning I didn't try to tell the boss I had saved the building. In that time he might just as well have decided to ask why I was not on that particular spot when the fire broke out. One never knew. I did a lot of different jobs in that hatchery and a few marks are with me still. At hatching time there were hundreds of trays to be washed manually with lye. The solution soaked the leg of my work pants. After some weeks I discovered a numb patch on my left leg. The superficial sensory nerves were evidently destroyed and they have never regenerated. Workers' chances; there was no compensation then.

Anyhow, I came off lucky. The underlying muscles were not injured. From that job I went to heaving one-hundred-and-fifty-pound feed sacks from trucks into the storeroom.

This went on fine for a while and was a great muscle builder. I was beginning to develop the shoulders of an ox. Then, all of a sudden, I began to malinger. I literally could not understand myself. Between bouts of work during which I was beginning to stagger under sacks I had once handled with ease, I crept off to hide in the dark shed, throwing myself in despair on the stacked grain. I was used to earning my pay in that hatchery. I could not do it any more. I literally could not do it. A doctor in that time was the last thing one thought about, he was next to the undertaker, and who had the money?

I struggled on. One morning I caught a glimpse of myself in a store window on a downtown street. Good God, my skin was grey! I could see its color even in the reflection from the window. I went back to my room and looked in the mirror. A kind of wan death's head looked back at me. No wonder the last tossed grain sack had nearly ruptured me.

I picked up the phone and called the boss at the hatchery. I explained carefully that there was an illness in the family and I would have to take a few days off. He was decent about it, but we were both noncommittal. Then I looked in the mirror again and promptly lay down on my cot. Something would have to be done. Even the effort to think was exhausting. I ran over the possibilities in my mind while all the time I never wanted to move again. My heart pounded and fluttered. But the kids went right on yelling next door. Basically what I needed was a graveyard. I put my hands over my ears to shut out the noise.

The next morning I went to the university dispensary, upon which I had some small claim because I had been endeavoring to take part-time work toward my long-lost degree. The doctor clocked my pulse at over one hundred. There was a slight tremor in my hands (it is with me still) and the blood tests showed severe secondary anemia. I was subjected to the rather clumsy hyperthyroid test of the time, but the results revealed no involvement of that nature. I could see the young doctor was puzzled.

At that moment there chanced to walk through the door a diagnostician of note in the medical community, a boistrous, loud-mouthed man whose avocation was the writing of occult fiction. Seizing the opportunity to consult so eminent a colleague, the dispensary physician beckoned him into his private quarters. As the man brushed past he gave me the passing glance bestowed upon stockyard cattle one does not expect to see again. The door they closed upon me was not the door of a modern consultation room. It was the door of an old nineteenth-century classroom. The hinges were loose and the door was as cracked as that in a workers' boarding house.

In short, the door was a fiction. The voice of the diagnostician would have penetrated oak. "Tuberculosis," the bellow came through the paneling. "Advanced. Not much chance. It's in his glands, that's why you didn't get it with the stethoscope. Prescription? All the milk he can drink. Maybe four quarts a day. But the chances of survival? No. Not much."

The door, I remember, shook to his voice. I was young. I wasn't interested in dying. I remember feeling the steady climbing of my pulse for which the dispensary doctor, after the departure of his visitor, gave me a prescription and some soothing words. Obviously he did not wish me to come back. I was simply a walking menace to the school. I thanked him and made my way to the street. I never saw him again. Something was wrong, all right, but I did not intend to be bulldozed off the scene by a specialist who diagnosed at a glance and who bellowed through a door knowing all the time you were poor and trembling with fear, and trying not to listen to his words.

I had a last card to play. Perhaps I believed in luck, then anyhow. It had favored me on that speeding mail train not many months before. I went home to my aunt's place and drank a quart of milk. Then I started to think. I wasn't coughing or spitting blood, but I knew glandular tuberculosis could in the end be equally devastating. I went out on the side porch under the honeysuckle vine. I stripped to my shorts and started sunning. I also knew I was running a temperature and in the days that followed I watched it follow the tubercular pattern. My weight had dropped from one hundred and seventy to one hundred and forty pounds. The worst of that time that I remember was the waiting. No one in my family expected me to survive: it was written on their faces.

In that day I had one bit of fortune which I will never forget. My uncle and aunt were childless. They belonged to an economic level which had been somewhat more secure, in the terms of that time, than our own. I was welcome in their home. I had been sleeping on a cot on their back porch since my father's death. Only a few nights previously I had started frantically out of sleep, clutching the edges of the cot as though I were still dozing on a freight-top runway. A whistle, blocks away, from a passing Rock Island locomotive, had penetrated my sleep. Long afterward I would hear once more the howl of that whistle opening up ahead, but that was years away.

With the modest financial help afforded by my uncle I sought examination by a skilled private physician. He listened carefully to my chest and heart, looked dispassionately at my thermometer reading.

"Been frightened, haven't you?" he questioned. I explained why.

"All right, listen to me," he said gently. "It is not glandular, that I can assure you."

"But the man said—" I started to protest.

"Never mind," he said, letting his brother physician slide into limbo without comment. He frowned doubtfully, listened again to my chest. "It's here, the râles, I mean, but faint, dubious, incipient." He tapped my chest and listened again. "You haven't been eating enough."

I said nothing.

"A bug has jumped you, but it isn't in your glands, it's in here. Sure, you've got to build up. Drink the milk, eat, but rest. Get it? Rest. You're not going to do anything for months. You will record temperatures and see me often."

I thought fleetingly of a girl I had known in high school, the sweetheart of a friend. It had begun in this way and after a year in a sanatorium she had simply faded away before the eyes of her broken-hearted parents.

"I'll try," I said. The office door closed quietly behind me.

In the end I fought the weight. I even cheated on myself by drinking water before I weighed in. I was just barely holding my own. Finally in midsummer the doctor spoke again. "You ought to get out of the heat. Somewhere high up. Colorado?"

"If I can manage," I said doubtfully.

It would be good, at least, to escape some of that circle of waiting eyes about me. To be fair, they had their reasons. Two of my uncle's brothers had died of the disease and he himself had had a well-nigh miraculous recovery.

My mother never expected to see me again. Among the Coreys—my mother's people—no one went anywhere except to die. From the small store of insurance money left by my father's death, she refused to contribute. Fortunately there was a little sum left directly to me. I withdrew it from the bank.

My aunt, a genteel little Victorian figure, suppressing her doubts, went with me. At Manitou, in the mountainous foothills below Pike's Peak, we rented a cabin. The treatment, as it was then conceived, was just to go on breathing while the lesion calcified. If it did you won, if it spread you lost. The high climate was supposed to help.

THE LIFE MACHINE

For a person as restless as myself the immobilization was terrifying. My mind raced endlessly upon itself. The fight with my weight continued. It held, it barely held. I used to wander down and sit in the square when my claustrophobia mounted beyond my ability to control it.

Far off over the horizon people were working, going to school, meeting girls, making their way in life. Engaged in this absurd warfare with my body, I was going nowhere. In that cabin I do not believe I had even a book to read. I existed and I thought; that was all. I was in my twenties and the watchers went on expectantly watching. Belatedly I do not blame them. After all, they were part of my heritage. It was also the atmosphere of the depression years. Nothing was supposed to turn out right and rarely did. Banks failed and people died. That was just the way of it. Even the bank robbers knew it, spent hastily, and lived by the code. But there I sat and unwillingly reviewed the past. It was scarcely edifying. Finally the money was all gone.

CHAPTER 3

The Running Man

WHILE I endured the months in the Colorado cabin, my mother, who had been offered a safe refuge in the home of her sister, quarreled and fought with everyone. Finally, in her own inelegant way of putting things, she had "skipped town" to work as a seamstress, domestic, or housekeeper upon farms. She was stone deaf. I admired her courage, but I also knew by then that she was paranoid, neurotic and unstable. What ensued on these various short-lived adventures I neither know to this day, nor wish to know.

It comes to me now in retrospect that I never saw my mother weep; it was her gift to make others suffer instead. She was an untutored, talented artist and she left me, if anything, a capacity for tremendous visual impressions just as my father, a one-time itinerant actor, had in that silenced household of the stone age —a house of gestures, of daylong facial contortion—produced for me the miracle of words when he came home. My mother had once been very beautiful. It is only thus that I can explain the fatal attraction that produced me. I have never known how my parents chanced to meet.

There will be those to say, in this mother-worshipping culture, that I am harsh, embittered. They will be quite wrong. Why should I be embittered? It is far too late. A month ago, after a passage of many years, I stood above her grave in a place called Wyuka. We, she and I, were close to being one now, lying like the skeletons of last year's leaves in a fence corner. And it was all nothing. Nothing, do you understand? All the pain, all the anguish. Nothing. We were, both of us, merely the debris life always leaves in its passing, like the maimed, discarded chicks in the hatchery trays—no more than that. For a little longer I would see and hear, but it was nothing, and to the world it would mean nothing.

I murmured to myself and tried to tell her this belatedly: Nothing, mama, nothing. Rest. You could never rest. That was your burden. But now, sleep. Soon I will join you, although, forgive me, not here. Neither of us then would rest. I will go far to lie down; the time draws on; it is unlikely that I will return. Now you will understand, I said, touching the October warmth of the gravestone. It was for nothing. It has taken me all my life to grasp this one fact.

I am, it is true, wandering out of time and place. This narrative is faltering. To tell the story of a life one is bound to linger above gravestones where memory blurs and doors can be pushed ajar, but never opened. Listen, or do not listen, it is all the same.

I am every man and no man, and will be so to the end. This is why I must tell the story as I may. Not for the nameless name upon the page, not for the trails behind me that faded or led nowhere, not for the rooms at nightfall where I slept from exhaustion or did not sleep at all, not for the confusion of where I was to go, or if I had a destiny recognizable by any star. No, in retrospect it was the loneliness of not knowing, not knowing at all.

I was a child of the early century, American man, if the term may still be tolerated. A creature molded of plains' dust and the seed of those who came west with the wagons. The names

Corey, Hollister, Appleton, McKee lie strewn in graveyards from New England to the broken sticks that rotted quickly on the Oregon trail. That ancient contingent, with a lost memory and a gene or two from the Indian, is underscored by the final German of my own name.

How, among all these wanderers, should I have absorbed a code by which to live? How should I have answered in turn to the restrained Puritan, and the long hatred of the beaten hunters? How should I have curbed the flaring rages of my maternal grandfather? How should—

But this I remember out of deepest childhood—I remember the mad Shepards as I heard the name whispered among my mother's people. I remember the pacing, the endless pacing of my parents after midnight, while I lay shivering in the cold bed and tried to understand the words that passed between my mother and my father.

Once, a small toddler, I climbed from bed and seized their hands, pleading wordlessly for sleep, for peace, peace. And surprisingly they relented, even my unfortunate mother. Terror, anxiety, ostracism, shame; I did not understand the words. I learned only the feelings they represent. I repeat, I am an American whose profession, even his life, is no more than a gambler's throw by the firelight of a western wagon.

What have I to do with the city in which I live? Why, far to the west, does my mind still leap to great windswept vistas of grass or the eternal snows of the Cascades? Why does the sight of wolves in cages cause me to avert my eyes?

I will tell you only because something like this was at war in the heart of every American at the final closing of the westward trails. One of the most vivid memories I retain from my young manhood is of the wagon ruts of the Oregon trail still visible on the unplowed short-grass prairie. They stretched half a mile in width and that was only yesterday. In his young years, my own father had carried a gun and remembered the gamblers at the green tables in the cow towns. I dream inexplicably at times of a gathering of wagons, of women in sunbonnets and black-garbed, bewhiskered men. Then I wake and the scene dissolves.

I have strayed from the Shepards. It was a name to fear but this I did not learn for a long time. I thought they were the people pictured in the family Bible, men with white beards and long crooks with which they guided sheep.

In the house, when my father was away and my mother's people came to visit, the Shepards were spoken of in whispers. They were the mad Shepards, I slowly gathered, and they lay somewhere in my line of descent. When I was recalcitrant the Shepards were spoken of and linked with my name.

In that house there was no peace, yet we loved each other fiercely. Perhaps the adults were so far on into the midcountry that mistakes were never rectifiable, flight disreputable. We were Americans of the middle border where the East was forgotten and the one great western road no longer crawled with wagons.

A silence had fallen. I was one of those born into that silence. The bison had perished; the Sioux no longer rode. Only the yellow dust of the cyclonic twisters still marched across the landscape. I knew the taste of that dust in my youth. I knew it in the days of the dust bowl. No matter how far I travel it will be a fading memory upon my tongue in the hour of my death. It is the taste of one dust only, the dust of a receding ice age.

So much for my mother, the mad Shepards, and the land, but this is not all, certainly not. Some say a child's basic character is formed by the time he is five. I can believe it, I who begged for peace at four and was never blessed for long by its presence.

The late W. H. Auden once said to me over a lonely little dinner in New York before he left America, "What public event do you remember first from childhood?" I suppose the massive old lion was in his way encouraging a shy man to speak. Being of the same age we concentrated heavily upon the subject.

"I think for me, the Titanic disaster," he ventured thoughtfully.

"Of course," I said. "That would be 1912. She was a British ship and you British have always been a sea people."

"And you?" he questioned, holding me with his seamed features that always gave him the aspect of a seer.

I dropped my gaze. Was it 1914? Was it Pancho Villa's raid into New Mexico in 1916? All westerners remembered that. We wandered momentarily among dead men and long-vanished events. Auden waited patiently.

"Well," I ventured, for it was a long-held personal secret, "It was an escape, just an escape from prison."

"Your own?" Auden asked with a trace of humor.

"No," I began, "it was the same year as the Titanic sinking. He blew the gates with nitroglycerin. I was five years old, like you." Then I paused, considering the time. "You are right," I admitted hesitantly. "I was already old enough to know one should flee from the universe but I did not know where to run." I identified with the man as I always had across the years. "We never made it," I added glumly, and shrugged. "You see, there was a warden, a prison, and a blizzard. Also there was an armed posse and a death." I could feel the same snow driving beside the window in New York. "We never made it," I repeated unconsciously.

Auden sighed and looked curiously at me. I knew he was examining the pronoun. "There are other things that constitute a child," I added hastily. "Sandpiles, for example. There was a lot of building being done then on our street. I used to spend hours turning over the gravel. Why, I wouldn't know. Finally I had a box of pretty stones and some fossils. I prospected for hours alone. It was like today in book stores, old book stores," I protested defensively.

Auden nodded in sympathy.

"I still can't tell what started it," I went on. "I was groping, I think, childishly into time, into the universe. It was to be my profession but I never understood in the least, not till much later. No other child on the block wasted his time like that. I have never understood my precise motivation, never. For actually I was retarded in the reading of clock time. Was it because, in the things found in the sand, I was already lost and wandering instinctively—amidst the debris of vanished eras?"

"Ah," Auden said kindly, "who knows these things?"

"Then there was the period of the gold crosses," I added. "Later, in another house, I had found a little bottle of liquid gilt my mother used on picture frames. I made some crosses, carefully whittled out of wood, and gilded them till they were gold. Then I placed them over an occasional dead bird I buried. Or, if I read of a tragic, heroic death like those of the war aces, I would put the clipping—I could read by then—into a little box and bury it with a gold cross to mark the spot. One day a mower in the empty lot beyond our backyard found the little cemetery and carried away all of my carefully carved crosses. I cried but I never told anyone. How could I? I had sought in my own small way to preserve the memory of what always in the end perishes: life and great deeds. I wonder what the man with the scythe did with my crosses. I wonder if they still exist."

"Yes, it was a child's effort against time," commented Auden. "And perhaps the archaeologist is just that child grown up."

It was time for Auden to go. We stood and exchanged polite amenities while he breathed in that heavy, sad way he had. "Write me at Oxford," he had said at the door. But then there was Austria and soon he was gone. Besides one does not annoy the great. Their burdens are too heavy. They listen kindly with their eyes far away.

After that dinner I was glumly despondent for days. Finally a rage possessed me, started unwittingly by that gentle, gifted man who was to die happily after a recitation of his magnificent verse. For nights I lay sleepless in a New York hotel room and all my memories in one gigantic catharsis were bad, spewed out of hell's mouth, invoked by that one dinner, that one question,

what do you remember first? My God, they were all firsts. My
brain was so scarred it was a miracle it had survived in any
fashion.

For example, I remembered for the first time a ruined farmhouse that I had stumbled upon in my solitary ramblings after
school. The road was one I had never taken before. Rain was
falling. Leaves lay thick on the abandoned road. Hesitantly I
approached and stood in the doorway. Plaster had collapsed
from the ceiling; wind mourned through the empty windows. I
crunched tentatively over shattered glass upon the floor. Papers
lay scattered about in wild disorder. Some looked like school
examination papers. I picked one up in curiosity, but this, my
own mature judgment tells me, no one will believe. The name
Eiseley was scrawled across the cover. I was too shocked even to
read the paper. No such family had ever been mentioned by
my parents. We had come from elsewhere. But here, in poverty
like our own, at the edge of town, had subsisted in this ruined
house a boy with my own name. Gingerly I picked up another
paper. There was the scrawled name again, not too unlike my
own rough signature. The date was what might have been expected in that tottering clapboard house. It read from the last
decade of the century before. They were gone, whoever they
were, and another Eiseley was tiptoeing through the ruined
house.

All that remained in a room that might in those days have
been called the parlor were two dice lying forlornly amidst the
plaster, forgotten at the owners' last exit. I picked up the pretty
cubes uncertainly in the growing sunset through the window,
and on impulse cast them. I did not know how adults played, I
merely cast and cast again, making up my own game as I played.
Sometimes I thought I won and murmured to myself as children will. Sometimes I thought I lost, but I liked the clicking
sound before I rolled the dice. For what stakes did I play, with
my childish mind gravely considering? I think I was too naive

for such wishes as money and fortune. I played, and here memory almost fails me. I think I played against the universe as the universe was represented by the wind, stirring papers on the plaster-strewn floor. I played against time, remembering my stolen crosses, I played for adventure and escape. Then, clutching the dice, but not the paper with my name, I fled frantically down the leaf-sodden unused road, never to return. One of the dice survives still in my desk drawer. The time is sixty years away.

I have said that, though almost ostracized, we loved each other fiercely there in the silent midcountry because there was nothing else to love, but was it true? Was the hour of departure nearing? My mother lavished affection upon me in her tigerish silent way, giving me cakes when I should have had bread, attempting protection when I was already learning without brothers the grimness and realities of the street.

There had been the time I had just encountered the neighborhood bully. His father's shoulder had been long distorted and rheumatic from the carrying of ice, and the elder son had just encountered the law and gone to prison. My antagonist had inherited his brother's status in the black Irish gang that I had heretofore succeeded in avoiding by journeying homeward from school through alleys and occasional thickets best known to me. But now brother replaced brother. We confronted each other on someone's lawn.

"Get down on your knees," he said contemptuously, knowing very well what was coming. He had left me no way out. At that moment I hit him most inexpertly in the face, whereupon he began very scientifically, as things go in childish circles, to cut me to ribbons. My nose went first.

But then came the rage, the utter fury, summoned up from a thousand home repressions, adrenalin pumped into me from my Viking grandfather, the throwback from the long ships, the berserk men who cared nothing for living when the mood came on them and they stormed the English towns. It comes to me now that the Irishman must have seen it in my eyes. By nature I was a quiet reclusive boy, but then I went utterly mad.

The smashed nose meant nothing, the scientific lefts and rights slicing up my features meant nothing. I went through them with body punches and my eyes. When I halted we were clear across the street and the boy was gone, running for home. Typically I, too, turned homeward but not for succor. All I wanted was access to the outside watertap to wash off the blood cascading down my face. This I proceeded to do with the stoical indifference of one who expected no help.

As I went about finishing my task, my mother, peering through the curtains, saw my face and promptly had hysterics. I turned away then. I always turned away. In the end, not too far distant, there would be an unbridgeable silence between us. Slowly I was leaving the world she knew and desperation marked her face.

I was old enough that I obeyed my father's injunction, reluctantly given out of his own pain. "Your mother is not responsible, son. Do not cross her. Do you understand?" He held me with his eyes, a man I loved, who could have taken the poor man's divorce, desertion, at any moment. The easy way out. He stayed for me. That was the simple reason. He stayed when his own closest relatives urged him to depart.

I cast down my eyes. "Yes, father," I promised, but I could not say for always. I think he knew it, but work and growing age were crushing him. We looked at each other in a blind despair.

I was like a rag doll upon whose frame skins were tightening in a distorted crippling sequence; the toddler begging for peace between his parents at midnight; the lad suppressing fury till he shook with it; the solitary with his books; the projected fugitive running desperately through the snows of 1912; the dice player in the ruined house of his own name. Who was he, really? The man, so the psychologists would say, who had to be shaped or found in five years' time. I was inarticulate but somewhere, far forward, I would meet the running man; the peace I begged for between my parents would, too often, leave me sleepless. There was another thing I could not name to Auden. The fact that I remember it at all reveals the beginning of adulthood and a sense of sin beyond my years.

To grow is a gain, an enlargement of life; is not this what they tell us? Yet it is also a departure. There is something lost that will not return. We moved one fall to Aurora, Nebraska, a sleepy country town near the Platte. A few boys gathered to watch the van unload. "Want to play?" ventured one. "Sure," I said. I followed them off over a rise to a creek bed. "We're making a cave in the bank," they explained. It was a great raw gaping hole obviously worked on by more than one generation of troglodytes. They giggled. "But first you've got to swear awful words. We'll all swear."

I was a silent boy, who went by reading. My father did not use these words. I was, in retrospect, a very funny little boy. I was so alone I did not know how to swear, but clamoring they taught me. I wanted to belong, to enter the troglodytes' existence. I shouted and mouthed the uncouth, unfamiliar words with the rest.

Mother was restless in the new environment, though again my father had wisely chosen a house at the edge of town. The population was primarily Scandinavian. She exercised arbitrary judgment. She drove good-natured, friendly boys away if they seemed big, and on the other hand encouraged slighter youngsters whom I had every reason to despise.

Finally, because it was farmland over which children roamed at will, mother's ability to keep track of my wide-ranging absences faltered. On one memorable occasion her driving, possessive restlessness passed out of bounds. She pursued us to a nearby pasture and in the rasping voice of deafness ordered me home.

My comrades of the fields stood watching. I was ten years old by then. I sensed my status in this gang was at stake. I refused to come. I had refused a parental order that was arbitrary and un-called for and, in addition, I was humiliated. My mother was behaving in the manner of a witch. She could not hear, she was violently gesticulating without dignity, and her dress was some-how appropriate to the occasion.

Slowly I turned and looked at my companions. Their faces could not be read. They simply waited, doubtless waited for me to break the apron strings that rested lightly and tolerably upon themselves. And so in the end I broke my father's injunction; I ran, and with me ran my childish companions, over fences, tumbling down haystacks, chuckling, with the witch, her hair flying, her clothing disarrayed, stumbling after. Escape, escape, the first stirrings of the running man. Miles of escape.

Of course she gave up. Of course she never caught us. Walk-ing home alone in the twilight I was bitterly ashamed. Ashamed for the violation of my promise to my father. Ashamed at what I had done to my savage and stone-deaf mother who could not grasp the fact that I had to make my way in a world unknown to her. Ashamed for the story that would penetrate the neighbor-hood. Ashamed for my own weakness. Ashamed, ashamed.

I do not remember a single teacher from that school, a single thing I learned there. Men were then drilling in a lot close to our house. I watched them every day. Finally they marched off. It was 1917. I was ten years old. I wanted to go. Either that or back to sleeping the troglodyte existence we had created in the cave bank. But never home, not ever. Even today, as though in a far-off crystal, I can see my running, gesticulating mother and her distorted features cursing us. And they laughed, you see, my companions. Perhaps I, in anxiety to belong, did also. That is what I could not tell Auden. Only an unutterable savagery, my savagery at myself, scrawls it once and once only on this page.

CHAPTER 4

The Desert

WHEN I returned in the fall the doctor listened to my breathing. "Holding," was all he said, "just holding. No hard labor, no school, and stay outdoors. The west would still be good for you."

"But the money," I thought. What was left of the two families was pretty well broken. An embittered and widowed sister had descended upon my uncle. He had his own problems, and I had no intention of being a contending rival for charity.

At this point a rumor about my illness must have passed among certain professors I knew at the University of Nebraska. One individual generously came to see me. "There is a very well-to-do woman in California whom I happen to know," this man said. "I have written to her about you. She has a vast undeveloped acreage over in the Mohave Desert east of Los Angeles. There is an artesian well on the property and a little cabin you may occupy. You would not have to do much but keep an eye on the place. The air is fine. The heat will be good for your affliction. I can arrange for your going. She will lend you a Model T for transport in the desert."

"All right," I assented reluctantly. There was no choice.

When I crawled out of my Pullman berth in Los Angeles, a kind-hearted traveler remarked solicitously that I looked ill and

gave me his card "in case of an emergency." "Thank you, sir,"
I said. "It is true I have been ill, but I hope the climate will
help me."

So it's here with me still, I thought despairingly. Even strangers can see it. I took a streetcar to Pasadena and waited, playing
little games of chance with a hotel acquaintance until my hostess chose to appear. It was a quiet place where birds sang in the
night. I suppose the gracious little hotel no longer exists. I was
young and I did not know I was looking at a dying way of life.

In due course my hostess and her chauffeur appeared—a man
introduced by the unlikely name of Nelson Goodcrown. They
were headed for the ranch. I had a brief impression of a short
brown-haired woman of distinct cultivation accompanied by a
muscular driver whose face was scarred and who, although
probably only a few years older than I, had had most of his
front teeth replaced with gold.

On the way to the ranch I was told what to expect. There
was a little cabin which I might inhabit that had once been
occupied by a man and his wife, now gone. The man had been
ill—here Mrs. Lockridge looked at me sidelong—and could not
leave the desert. Wisdom prevented me from asking what became of him. Some distance from this building was her own
private home, arranged in the best Spanish style. Farther out
was a guest house built in the sagebrush but within walking
distance of the house. It had large windows and was light and
airy but basically it was only a bedroom. There was no way to
cook there.

At a point on the road close to Death Valley, the huge Pierce-
Arrow swung aside upon a trail that ran off first into sage, and
then angled away across a huge dead lake basin as smooth as a
dance floor. After some five miles on the lakebed—a remnant
of ice age times—we reached the estate. The most welcome
sound in the world was the tinkle of water falling from an
artesian well into a huge concrete basin. The well was not yet
harnessed to any constructive work, but it alone must have cost

a fortune to dig. I wondered fleetingly how long it would be before the water table might be depleted. Miles to the west were great alfalfa fields similarly watered.

At a picnic supper drawn from a basket, with plentiful thermos jugs of coffee, Mrs. Lockridge explained that she was of a mind to raise turkeys on the ranch, say one thousand to begin with—again the curious sidelong glance. Not now, not immediately, of course, but in a few months if my health sustained me.

"You will have nothing to do but be a caretaker," my professor's words ran through my mind. Now, or so it seemed, I was shortly to be attending one thousand turkeys or find myself a most ungrateful guest. I shuddered. We had custom-raised turkey chicks at the hatchery, delicate birds if not treated properly. I nodded uneasily to Mrs. Lockridge.

"You can help Nelson build fences for a few days," she said. "Then I will have to go back to San Diego."

"Yes ma'am," I said. The brawny Nelson grinned at me.

"In the morning," he said. "That is your house over there." He gestured toward the little cabin.

Night had fallen, sprinkled with desert stars. The dead lake-bed in the distance glowed like a moon landscape. I picked up my things, took the key I was given, and trudged through the sand to the little unpainted shack. I did not go in, not then. That could wait until they all had gone. Instead, I crossed to the guest house and sat a long time on the stoop watching the magnified stars of the Mohave circle the horizon. There I fell asleep. I had no wish on this first night to sleep in the cabin of a man who had died of the thing I was still carrying.

When I roused myself in the dawn the wind had begun to blow. On that desert it blew perpetually until evening. It was hard to hear people's voices. One was always turning one's head to make sure one was not being hailed. Nelson and I labored for several days at a long section of fence, burning our bodies black in the desert sun. He had the easy good nature of a giant cat, not a developed muscle man who had to work at conditioning himself. Nelson by contrast was a complete natural.

Again, after Mrs. Lockridge's departure, we played poker in the evening by the light of a Coleman lantern. Slowly in that silence I learned that Nelson had been born in Detroit. I came to accept the fact that his strange-sounding upper-class name of Goodcrown was genuine. He had been an enforcer in the back alleys of the city. Finally the police, with the best of reasons, had suggested unofficially that he seek a career elsewhere. Detroit would not be healthy after a certain date. Nelson was not soft. The advice must have had a note of finality.

I studied covertly the scar tissue across his cheek. He had spent an impoverished period with a carnival, ripping telephone books apart, hoisting huge weights, taking on all comers in the various little towns frequented by the traveling show. Beyond these stray confidences Nelson never ventured. He, too, was an emergent out of wandering America. I never learned why he came to Los Angeles or how he had become the chauffeur of a wealthy lady.

He was not highly educated and he had, I discovered, some peculiar word blockages. The word "mirror" was one of them. With the frustrated eagerness to teach that besets young academics, I tried gently several times to correct this impediment. "Say 'mirror,' Nelson," I coached, and repeated it, emphasizing the pronunciation slowly.

"Maurer," dutifully repeated the great beast in a sleepy snarl like the Metro-Goldwyn-Mayer lion. I sighed. Who was I to press the virtues of education upon Nelson Goodcrown? He yawned, stretched his magnificent body, tossed the cards aside, and padded off to bed. I say padded advisedly. I always thought of him as a huge cat and just as dangerous. Nevertheless the time would come when he would need a "pronouncer."

Life at the ranch was without routine. It was early spring and incredible flowers sprinkled the thorny landscape of Joshua trees. There were long weeks in which the constant wind was maddening, the heat intolerable. Sometimes I wandered by the shores of the dead lake that had once been fed when there was glacial water in the distant mountains. Sometimes mirages danced in the heat waves. Mostly I was alone. Mrs. Lockridge had a home on the coast, and Nelson was kept busy driving her about on business whose purpose I never learned. On occasion I took the Ford and drove some thirty miles for supplies. The town I visited had nothing to offer in the way of entertainment. Besides, I was there to get the benefit of the desert. Lonely though I was, the place was doing me good. My color was better than when I had emerged from the Pullman berth. My weight, in the open air, was creeping upward; I was no longer immobilized. Sporadically, when Nelson appeared, we labored on the fence that was presumably intended to control the thousand turkeys of Mrs. Lockridge's dream, but her business affairs were multiple and devious. No more was said. I walked and slowly blackened in the Mohave sun. Lizards and pack rats were my common acquaintances.

The rats had known the moment I had settled into the little cabin. They were there before me. While I was unpacking supplies from the Ford pickup they stole a loaf of bread from the box of groceries in the truck bed. They had meddled among the few remaining possessions of my predecessor. For reasons best known to themselves they had chopped a whole fish line into neat lengths which they had then abandoned. At night they romped and squealed in the little inaccessible attic overhead.

One night I fought a desperate battle to retain the padlock to the back door. I had unlocked it and left it lying for a moment. One of the trade rats seized this opportunity. When I turned back the lock had vanished. I heard a clattering behind the cupboard. My flashlight caught the reflection of the little red-rimmed eyes of the rat. I managed with the aid of a stick to spear the padlock and recover it before it vanished forever into the mysterious realm of the traders who seized upon everything portable and left sticks in its place. Their boldness was incredible. I obtained the keys to the guest house which had been so planned that they could take no advantage of it. Even so, they snatched my glasses, which are probably to this day ornamenting a rat midden beneath a clump of cactus. I had to send home for another prescription.

During that same week of fence-laying, Mrs. Lockridge had driven off upon some errand of her own. Nelson and I had laid more fence, but evening had fallen without her return. Nelson chafed. "She'll get lost if she's not already," he grumbled.

The lake basin was particularly treacherous at night because landmarks were low and inconspicuous and had to be memorized. One's headlights merely wandered aimlessly over miles of the glowing lake floor. After supper we got in the Ford beside my unused cabin and waited upon the rise, watching the lake.

Miles off we saw the hesitant, wandering approach of the big car. "She's losing herself," affirmed Nelson. We drove out on the lakebed and hastened forward to meet her before she became confused. To our surprise the weak lights of the Ford came close to creating a catastrophe. Alone in the Pierce-Arrow, for whatever reason known to herself, Mrs. Lockridge panicked at sight of us. The car began to swing wildly in great circles. She no longer knew what she was doing.

"She's scared," growled Nelson, to whom this was the ultimate ignominy. "She's yellow, yellow, yellow." He pounded his hands on the wheel and started to step up our speed.

"Wait, Nelson, wait," I pleaded. "Sometimes people have what is called a phobia, something they can't control. She doesn't recognize us. She thinks we're after her. Show her. Give her time to get used to us."

Those were the old days before hippies, dune buggies, and violence had spread across the valley. But still she feared, she feared something, whether shadowy or real. In a flash I saw the reason for the giant chauffeur.

My words had some effect on the muttering, cursing Nelson.

THE DESERT

We approached very slowly. We honked encouragement. Once we stopped and I stood in front of our car lights waving and hoping to be identified. Slowly the frantic circling ceased. I believe, if she could have found the trail back to the highway, she would have fled, but she was beyond that now, helpless. Slowly, carefully, we drew up to her, calling as we did so.

I did not wish to be the witness to her embarrassment. I left Nelson to go across and drive her home, while I followed in the Ford. At the cabin I turned aside and parked. I did not go up to the big house. Of what was said I have no knowledge, but a fey sense of oncoming trouble pursued me. Mrs. Lockridge was a genteel woman with genteel intellectual friends. I had met a few whom she had invited up to the ranch, bringing a cook with her to see that her guests were made comfortable. Beneath it all, however, she was haunted by a great fear. Whether it was grounded in real events or was psychological in origin, I was never to learn, but someone in the valley had confided that she once had been severely beaten. Why, or by whom, I never learned.

Shortly after that evening's episode Nelson told me that Mrs. Lockridge had given us both a couple of days off. We would take her down to San Diego and leave the car with her. Then, exulted Nelson, we would go to Tia Juana. I didn't mind relief from the loneliness of my desert monastery, I didn't mind a glimpse of Tia Juana. What did give me pause, as a man slowly recovering from a dangerous illness, was my companion, the big, grinning cat who was purring about me a little too assiduously. If I was to intrude into the drinking spot which in those prohibition days was the Mecca of the Pacific fleet and the off-scourings of the whole southwest, there were certain inherent dangers in Nelson's company. He was a man born wicked, with the body to sustain, not restrain, his appetites.

After all, however, I was not indifferent to adventure; it was just that Nelson's laughter, as an ex-enforcer, was a little too hearty. I hid my fears. "Let's go," I said. It would be my first time over the border. The details are gone. A bus deposited us there in the late afternoon.

Nelson did not bother with the famous block-long bar. He walked into a store and bought a pint of whisky. Then, with the conditioned reflex of every American male of that period, he walked a few feet up the nearest alley. "Here," he gestured. "The first drink is yours."

I upended the bottle, held it a little longer than necessary, took a brief swallow, and handed it back to Nelson. Only a man with an iron constitution could have done what I saw him do. He let the entire pint of raw liquor blaze down his throat in pratically a single gulp. The bottle sagged out of his fingers. He drew a hand across his mouth and uttered one trailed-off word, "Whee'ee." "Keep this," he added dimly.

He handed me his wallet. I pocketed it unwillingly and trailed along. He was moving by instinct now. He knew what it meant to get rolled. We traveled up a street where Nelson had once met a Mexican girl who had tried to persuade him to marry her so she could get across the border into the states.

We passed a bakery. Nelson saw a pretty face inside and lunged in, with me clinging to his coat. It wasn't the girl or the sort of neighborhood he was looking for. Two girls dodged behind the counter, both giggling and appalled. The great beast was in rut. I was no brawler from the Detroit waterfront, that was for sure.

"Nelson," I bawled in his ear. "Let's get out of here before the cops take us in." Nelson must have heard this many times in the course of his life. He reeled but followed. Night was falling. I led him with such enticements as I could concoct well out of town on pocked roads.

Why hadn't I simply left him to stumble about in the dark street? God knows it was like trying to lead a Bengal tiger safely through a tough town and out again with no one dead or jailed, including myself. Still, he had trusted me in that pathetic gesture of giving me his wallet. This was our sorry party, under the street lights of a Mexican town. My introduction to sin.

It was there, in retrospect, that I first began to feel the power of words. "Nelson," I discoursed under a street lamp, watching the moments run, "we are Americans. We, you and I, are not going to get involved with cheap women. We are going to be able to look Mrs. Lockridge"—I threw in her name for its financial effect—"in the face tomorrow." I was either supremely eloquent in ways I can no longer remember, or Nelson was supremely drunk. Perhaps it was a combination of both. I gave all I had on the subject of patriotism forty years ago to one listener under a street light in Mexico. Only a brass band would have made it better. I have never equaled that moment since.

In some odd youthful way I had become involved because I had been trusted. I could not leave Nelson and if I led him where he wanted to go there would be knives, violence, and possibly, if we survived the night, jail. Nelson alternately wanted women and trouble. In his condition the trouble outweighed the women. I wanted none of it. I wished I were back in the desert starlight even if the pack rats did play chess with my belongings.

Back and forth we staggered as I alternately bore up Nelson's huge carcass and declaimed upon the manifest destiny of America of which he, Nelson Goodcrown, was a most notable example. Hours passed. We found ourselves crouched in a lot with a few other undesirables near the International fence, now closed. Nelson had vomited and lay groaning on the ground.

A Mexican policeman approached, polite and efficient. "This is your friend?" he asked. "You are with him, no? You will see to him?"

"I will see to him," I said. By then I had learned the border closing was a polite fiction. There was a hole in the fence up the road, through which the night people crawled. There were buses to take the sick and walking wounded back to San Diego.

"Nelson," I said, prodding him back to consciousness. "The

boss," I bellowed in his ear again, "the boss will be waiting for us in the morning."

"Christ," groaned the repentant Nelson. "I thought it would be you lying here, I wanted to see you drunk and crawling."

"Nelson," I repeated, helping him up, "the cops have been here. They'll be back. We've got to get through the fence and get on the bus. Come along now. We'll get coffee in San Diego." Somehow I got him through the fence and onto the bus. He was beginning to doze.

Up the road two officials boarded the bus to check citizenship. As they came down the aisle I whispered fiercely to Nelson, "Say American, Nelson." They reached us. "Maurercan," emerged desperately from Nelson's lips. It was a combination of trying for the word mirror and the place for natural functions. The officer sniffed, grinned, looked at me, and passed on. That one night had impressed me with such responsibilities that I have never been quite able to shed them since. I had brought my great purring tiger safely through sin city even if he was now a trifle bedraggled.

But we were not quite home. The coffee when we reached San Diego aroused Nelson to consciousness. He became fascinated with the legs of the waitress at the counter and began to lean heavily across it. We were back in the states and anyhow it was too late for another speech. "Nelson," I rapped sharply. "The cops, get it, remember what happened in Detroit. You've got a good job here. Keep it." Slowly I dragged him upright. We went across the street to a dingy hotel where I secured, under the dubious eye of the desk clerk, room twelve and a half, a euphemism for thirteen. Fear of the number may have its point, as I was to discover later.

Nelson became maniacally active once more. He wanted to show me he could rip the phone off the wall in one tug. "I believe you, Nelson," I said solemnly, as indeed I did, "but then we would have no way to call Mrs. Lockridge in the morning." Finally Nelson compromised. He unscrewed the mouthpiece, which I carefully made off with. Then we plotted to steal a worthless floor rug which Nelson was to wrap around himself when we left the hotel in the morning. With this arranged, Nelson climbed into bed and went soundly to sleep. I got in beside him and never slept till dawn. Looking back I sometimes think it was the most magnificent achievement of my life. I know now what the great animal trainers must feel.

In the morning I gave Nelson his wallet. He eyed it a little doubtfully. "Remember," I said cautiously, "you asked me to keep it for you. Everything is there." I did not elaborate for the simple reason that I had never looked in the wallet. He could have had five dollars in it, or five thousand. He took it silently but I could see something had gone from our relationship.

I supplied the missing telephone mouthpiece and he phoned Mrs. Lockridge. We headed back for the ranch. They dropped me there.

I spent another month of solitude in the desert. Then one night, though I only learned about it later, there was a drinking party in the kitchen of the big house. The men involved were part of an indeterminable underworld that seemed to circulate far out in Mrs. Lockridge's shadow. In a drunken moment someone had fired a shotgun through the ceiling.

Slowly it began to filter back to me, through a friendly rancher, that Mrs. Lockridge held me responsible for this vandalism. I never learned whether she honestly had been made to believe it by the men involved or was simply afraid to accept any other version of the event. The night on the lake bed when she was lost hinted of motivations I would never understand.

It was all ridiculous, of course. She knew I possessed no shotgun and that this strange group of men had keys and were known to her—had, in fact, access to her house. If Nelson had been one of the party, as I suspected, I never succeeded in finding out.

It was time to effect a polite escape and I found one. Evidently there were others who did not wish me on the property for their own purposes. I was in reasonable health and I intended to stay that way. October is a traveling month for both birds and men.

Some months later I met the Good Samaritan who had obtained the post for me. He asked me how I had fared. I said I felt better now and thanked him. "Mrs. Lockridge," he said glowering, "thinks you are mad. She wrote me that you had fired a shotgun through her roof." He was a man who liked wealth and automatically sided with it. Now he was furious. In some way he, too, was caught up in my mysterious offense. I shrugged. There was nothing to say that would have made the slightest difference. Benefactors, I had learned, were not always improved by their benefactions. That was the last time I saw him. I had been spared, it seemed, the care of one thousand turkeys. That, too, should be credited to room twelve and one-half in a shabby San Diego hotel.

CHAPTER 5

The Trap

THE râles in my chest were gone. My weight was approaching normal but if this was true, another disease gnawed at my vitals. A yearlong immobility, even my enforced wary care of myself in the Mohave, had left me savage, restless, at odds with my environment. I tried, through university extension courses, to overcome deficiencies and graduate. All failed. I prowled about like an animal. Suddenly, I vanished again. Always, as though it lingered in my blood, the ways, however wandering, lay west, not east. I remember from a train top the oak leaves turning red in Missouri, then the desert, California, the desert again. As in the case of all drifters, time was fading from my consciousness.

Across the whole southwest in the years before the thruways, the railroads climbed, descended, and wound through unimproved mesquite and gravel. Sometimes there was more gravel, sometimes more mesquite. Dusty roads ran aimlessly away into vast distances without water. At one roadside there was a sign that read: "Jobless men keep going. We can't take care of our own." There were many such signs across America in those years.

If you turn backward, I thought, grimly studying the sign, you will find only shutters banging in the wind, smashed windowpanes, the eyes of strangers, or, worse, the eyes of those who know you so thoroughly they wish never to see you again. When the thought struck me that I still might be welcome in my uncle's house, if only temporarily, I was huddled against an adobe wall in some nameless little desert town at nightfall. My teeth chattered in the cold of the high desert. I had no blanket and again no money. Sleep was impossible. In the morning I would have to search for food.

The place was a trap. The railroad ran through a concavity in the desert which contained the town. The police watched the two ends of the bottleneck. The town cops had every stranger spotted and merely waited, either for him to leave, which was difficult, or for him to steal, which could prove utterly disastrous.

The hungry and the beggars, of whom there were many in those years, watched the police and speculated. The police, in their turn, watched the hungry. It was a long chess game with generally but one ending because of heat and hunger and the railroad police at the bottleneck.

If you tried the desert you could either die unpleasantly or come back and commit a crime, thus qualifying for social assistance behind bars. The rules were very simple in that place, the world reduced to manageable proportions. The citizens didn't count. All that mattered were the watchers who ate and slept, and the watched who were unsuccessful at either.

The problem was to stay free and buzz out of the fly bottle. I did not know Wittgenstein in those years, nor the higher metaphysics. I was mercifully spared the knowledge that the real fly bottle was the world. There was only this particular small bottle to which circumstance had brought me. In the morning, if I had not died of pneumonia, there was time to seek companionship and a way out. On the back street of a Mexican slum avoided by the watchers, I would be reasonably safe. I continued to huddle inconspicuously beside the wall or pace in the shadows for warmth.

THE TRAP

In the morning the sun burst over the desert with an intensity to match the cold of the desert night. By then there were other vagrants at the wall. One said to me in desperation as we both eyed a bottle of milk which had just been deposited on a doorstep, "Suppose we . . ."

"Not here," I said. "It won't work. Believe me, it won't work."

The man smiled a little wanly. "I was just thinking," he said wistfully. "I used to be a mechanic. There's a car back up this street. A little wire fixin' and we could go right out of here."

I looked in his eyes, hesitating. He wasn't a bad man. He was just desperate. It was a moon landscape that made men desperate. One had to be desperate to live.

"Let's go up to the other end of town," I said to break the tension. "There must be some way out of here, otherwise they'd fill that jail in no time."

We climbed up to the cratered rim that circled the town and the railroad yards. After a long hungry wait at the end of the pass a train pulled by us. It was enormously long, but on the tender beside a young brakeman stood a dark-suited figure with the unmistakable air of the law about him. That line had a bad reputation among experienced drifters who pass such things along. There might be a division point dominated by some particularly ruthless individual who was capable of occasional killings just inside an easily stretchable law of trespass. It could happen at night. Some railroads issued strong orders to keep their freights clean of riders and did not recognize a depression.

Sometimes officials did not care to ask how the job was done. A dead man in a desert doesn't stay around very long. In that day men who left home to seek work or relieve family expense might never be heard of again. What happened to them was anybody's guess. No doubt some chose their own disappearance. It was not always so. And there were the bloody, Goddamned accidents. I ought to know. I carry my own scars.

"That dick gets off down the line aways," one man ventured

to me. "Has a car waiting for him, but, man, he's tough. They all say it."

We stood in silent frustration watching the climbing train on the grade. Suddenly I saw something. The brakeman was standing a little back of the black-clothed, gun-strapped law. In his light overalls and shirt the brakie was very visible and he was waving, waving surreptitiously with an unmistakable scooping motion behind his back. It might be a trap but we were desperate. Besides, the brakeman, as he passed, had looked young and friendly and pleasant-faced. We let a good half-mile of cars go by. Then we ran stumbling up the grade and swung aboard between two refrigerator cars. The sun beat mercilessly upon us. We didn't dare show our heads above the runway. We didn't know whether the law was gone. Finally, far up ahead, was a faint hesitation that traveled down the line of cars. We poised with eyes fixed on the grade if anyone challenged us. No one did. The train picked up speed. Riding between cars is not pleasant. Finally we became aware of the brakie's genial young face peering down at us.

"C'mon up," he gestured. We crawled up the side ladder and stood beside him. The black menace with the gun was gone. We followed our friend two car lengths down the runway. "Here," he said, grinning, and up-ended a reefer trapdoor. We peered in. There was a big, melting slab of ice lying there, sufficient both to sit on and to quench our thirst by chipping with a jackknife. "I'll see you to the next division," he said. "The worst will be over then. Don't be scared. I'll lower the trap a bit to shut out the sun and keep you hidden. Don't show yourselves on top. Lie low, lie low, and you'll soon be out of that bastard's jurisdiction." He smiled again. He was about my own age.

"Thanks," I said. "I won't forget." Crossing a desert on an ice cake? Of course I wouldn't forget. I have said he was about my own age. I just escape remembering his features and, of course, he is still young in my mind. It is I who have aged. I hope he remembers. I hope life was good to him. I suppose he helped others. I hope the man with the gun never cost him his job. It was a hard time to lose a job and that was a hard road. He was taking chances and didn't seem to care, though he was proficient at it. I wish I had said—I wish—but that was the child sense in me, time going away. I would like to have given him one of my gold crosses.

The food didn't matter, that could always wait. It was the ice, you see. I could have cried and embraced him while we merely grinned at each other. The temperature in that place could run one hundred and twenty degrees. Lost men have died in that desert with canteens. And we had no canteens. Later, much later, on good roads, I have passed that way in cars, but never without gallon cans of water in the trunk, and a care for the map. Oh, I remember that desert very well. I have an enormous respect for deserts, I have worked in them many times. I know them. I respect their inhabitants and their solitudes, but still I distrust them. One cannot expect to find an ice cake in a desert twice, nor have ice for a companion for four hundred miles.

The rest is almost forgotten with the years, the grim, long windings through the Rockies, viewing the scenery from side doors or pulling a bandana over one's face and being careful to duck low in smoky tunnels. People dropped away as the mood or hunger struck them. Others ran and climbed aboard. Here and there in small towns one tried to strike a bakery in the early morning and offer to sweep it out. An hour's work, if one was accepted, might be rewarded with a sack of stale rolls from yesterday's holdovers. Fair enough. One lived. And some bakers were kinder than others, particularly in places where the homeless did not swarm.

It was a time. Leave it at that. The young, the middle-aged, the old, even a few case-hardened women. We lay like windrows of leaves on sandbars beside the Union Pacific, the Rock Island, the Santa Fe, the Katy. At night our fires winked like the bivouacs of armies. We rode the empty fruit trains coming through Needles into the sand hell of the Mohave. Railroad detectives blackjacked us or turned aside in fear of numbers. We gathered like descending birds in spite of all obstacles. Like birds, some of us died because we were old and we perished, unnoticed, of cold in the high Sierras or we slipped under the wheels of freights in moments of exhaustion. If found, what remained was buried in nameless graves along the track. Cheap liquor killed us; occasionally we died by the gun and so did the railroad detectives, pushing their luck too far with sullen unknown men in the night on swaying car tops.

It was a time of violence, a time of hate, a time of sharing, a time of hunger. It was all that every human generation believes it has encountered for the very first time in human history. Life is a journey and eventually a death. Mine was no different than those others. But this is in retrospect. At that time I merely lived, and each day, each night, was different.

Somewhere in Colorado I lay over and waited for a night train, catching it on the fly after a race from behind a lighted signboard. We bored fast and furiously, into the plains. At a water tank two men climbed the first blind. We were crowded. I moved to an iron ladder behind the engine tender to accommodate them. We were running at a frightful speed and I was taking most of the jouncing. Kansas was almost open country in those days, with little to fear. Men got on, men got off. I was determined to ride that express into another dawn.

I was drugged with fatigue without knowing it. I had a vise-like grip on the tender ladder but I was directly above the wheels. The pounding had dimmed into one long interminable roar. Still I clung. I was headed home. For the moment I did not realize how little of home remained. I was starting to fall asleep in the most dangerous spot in the world, the spot from which others had never wakened. The hands, I thought dimly; I shifted them and clenched once more. I called out in the solitude. I think it was that which roused me. I was alone, no one answered me. The men I had ridden with were gone, dropped off stations ago. I drowsed again and spoke to phantoms in the black dark. My hands still held. Slowly, even as I slid downward, something in my body, something divorced from my groggy conscious mind, screamed in protest. "Wake up. In five more seconds you'll be shredded to bits. For Christ's sake, man, there is nobody there, you're talking to yourself. You don't care anymore, but I do. I'm your body. Straighten up, do something, anything. We're going to die."

How many hours, how many miles, how much sleep loss can a man take? What mania had possessed me to outride relays of riders in the most dangerous, muscle-jerking spot on a locomotive? I was fully awake now, but too weak even to crawl over the tender on to the coal, or back to the safety of the blind. I shook my head, tightened my cramped fingers, and waited for the next stop. It wouldn't happen again, it wouldn't, it wouldn't. I had seen one death of this kind. I didn't intend to make a second. We flashed over a bridge I thought I dimly recognized. We slowed for a station. I waited till we stopped. Never mind the cops. They could have me. I descended and stood silently beside the tender. Only a tired switchman appeared as the fireman high on the tender took on water.

"Where are we?" I asked the switchman in the dark.

"Kansas City," he said, and then as an afterthought, peering into my alarmed face, "Not the station. Not yet."

"Honest?" I said.

"You ought to be careful, son. You could kill yourself, you know." He meant it kindly. "Where you goin' in such a hurry?"

Good God, I thought. I've crossed a state almost in one run. You fool, you utter fool. He's right. You could be dead.

Aloud I said, "I'm going home." The man shrugged wearily, as if he had heard the words a thousand times and disbelieved them a thousand more. "Yes," I reiterated, "I'm going home, but not on this train. I'm dead beat, Mister. Where can I hole up?"

"There's a place up the tracks," the switchman pointed. "See that little fire? Somebody's generally there. You might find somebody with a little food who'd share. In case you don't and since you didn't ask, here's a quarter." He seemed embarrassed, even in the dark. He wanted badly to believe I was going home. A home man himself, it meant something to him. Maybe there was a kid.

An element of respectability reasserted itself in me in response. "You sure you can spare it? You've been decent, awfully decent. That train," I gestured, "has darn near been the death of me. You're right, I won't argue any. I rode her too long. I caught myself slipping, right there on that ladder. I was nodding off to sleep. I won't ride again tonight."

The man visibly shuddered. "Go up there to the light, son. It's almost daybreak. Nobody will bother you, get some sleep. My God, you must have set some kind of record." For the first time he grinned. "You really are in a hurry. I believe you."

I touched his shoulder in a kind of salute. "Thanks again," I said and stumbled sleepily toward the light. I never got there. I knew I wouldn't, but I had to keep face with my friend the switchman. Out of his sight I staggered off the rails into some underbrush and fell immediately to sleep.

But give me this: with my last conscious act before I collapsed I stumbled off the tracks. I once knew two men so tired they lay down upon the tracks. It was their last sleep. I know the place, but it will remain nameless. The wind and the night know the secret. But this one act I had sense enough to perform before I dropped: get off the rails. It was the last I knew. I never reached the light, the food, or the camp. Not that night. Among the hard stalks in a sunflower thicket I did not even dream. And now—now, of course, I would give anything for sleep.

CHAPTER 6

Toads and Men

I SLEPT dreamlessly, hidden in the sunflower thicket, until late afternoon. Then I walked, bruised and stiffened, to the hobo camp beside the tracks. I had a horned toad in a sack. There were railroad police, the switchman had told me, but none in that spot, and the air came out of a little gully beside the track and it was cool. Hours would pass before there was a night freight. Over on the horizon hung the towers of the city, white in the faraway sunlight. It was the third year of the Great Depression in the railroad yards of Kansas City.

"Where ya headin'?" a squatting youth asked. He pecked a cinder at a keg of sleeper spikes as I crouched beside him in the shade.

"East," I said, and knew that was enough and that none of us was headed differently.

"There's nothin' there," he said, and pecked viciously with another cinder. The barrel shook a little.

"A man's got to eat," I ventured mildly. "Take that desert, now, no more of that for me. You get off one of those trains in a town and there's so many of you that they load you right back on with shotguns and ship you out. Don't want to feed you. Too many people, that's what."

The youth kept up a surly battering at the barrel. A man with a white prison face was more friendly. "I'm headin' for home," he said. Then, a little doubtfully, "I guess they're still livin' there. Things are kind of changed." He looked at his hands. "My stretch was up a week ago in L.A. I've got this far."

"Yes," I said. "So have I." But I didn't say to where. A line from Sandburg went idly through my head: "All the coaches will be scrap and rust and the people ashes."

Another little group of wanderers came down the tracks and sat hesitantly among us. Whether they sat or squatted, they all held their hands loosely across their knees as though they were waiting unconsciously for either one of two things to happen: either a tool or a stone to be thrust into their fists. Until that was done they would crouch here, or in a hundred similar places, waiting until the city on the horizon, in its own good time, called for them again. I sat there and flexed my hands like the rest. I threw a stone at the barrel.

"The bastards," said one, "the God-damned bastards." He said it ritualistically, without heat, like a stone-age man in evil luck cursing unfeeling gods.

"I think it's done," said another. "I honest to God think it's done. It'll just go on till everybody is out here eatin' slum. Then maybe we can start over." A little approving mutter ran along the right-of-way, but the man with the prison pallor drew off a little and huddled down into his coat.

"I'm glad to be out," he said, "just right out here breathin' air." His brown, hurt eyes traveled over several faces and came to rest on mine. "Ain't nobody can fix up things havin' to do with people," he said, suddenly giving his attention to me. "Ain't nobody can do that. People got no knack that way. People can make railroads, people can build all them buildin's"—he gestured toward the skyline—"but they can't figure *themselves,* and they're always mean when they git tryin'. Don't let 'em fool you, kid."

He paused suddenly, stricken with thoughts too big for him.

"I'm goin' home," he said, "I'm gittin' back to my old woman; the rest can take care of itself. T'hell with it."

"T'hell with it," came the ready respectful chorus of the group.

The man got up slowly under some enormous tension and moved steadily off down the track in the hot sun. No one called after him, and even the thrower at the barrel was quiet for a little while.

It was just then my toad began to rustle. I had been squatting there with the sack in my hand, and some of the men had been eyeing it speculatively as representing food. But right then that fool toad had to rustle. Maybe he didn't like the heat and was getting tired of being shut in. Anyhow he scratched till I couldn't deny that there was something alive in the sack.

"For Christ's sake," said an old man with blinking, cinder-beaten eyes. "What you got, huh?"

I let the thing crawl on my hand. It quieted the animal, and when I rubbed his sides alternately with a finger, he would puff and tilt toward the finger and all his spines would show. He was a fine sight.

"I got him this side the mountains," I said. "I had to walk ten miles cross country after being kicked off a freight. He was in the sand."

"You oughta stomp on him," said the cinder thrower after a brief, bitter glance. "Them things might be poison."

"The hell," said another. "Them things is all right. They been there a long time in that sand. They won't hurt you. Lucky they are, like a four-leaf clover, that's the reason he's carryin' it. For luck now."

"Well, no—" I started to say.

"I know," said another more kindly. "It's nice to have some-thin' to travel with. I had a pal once had a dog—little fox terrier—he'd even grab a train with that dog under his arm. They got him for a stick-up in the Springs. I don't know what happened to the dog. This is better, you can carry him in your pocket."

"I don't know," I said. "I liked him so much I put him in the sack. He puffs out fine when you rub him."

"He'll die anyhow," said the cinder expert, beating a rolling tattoo on the battered barrel. Its staves sprung, the barrel collapsed slowly, and its contents spilled down the embankment.

"I wonder where all them toads come from," said the man with the burned eyes, lighting a pipe. "It makes you wonder about things—those spikes on his head now."

"Where he comes from? Aah," said the cinder thrower, with fine illogic, "he's a God-damned toad, that's what he is. You can't eat a toad." He withered me with a glance.

"There might be a medicine in him," said a man with a quick, constant sniff. "The boys in the carnival used to say slit 'em right up the belly and when you find—"

I gave the toad one more rub with my finger and when he tilted, I slid him slowly back into the sack.

The man with the pipe shook his head. "Men or toads—they all had to come from somewhere," he grumbled. "Calling 'em toads or calling 'em men don't answer that. Mark my words, something had to have a hand in making us."

"Yeah," philosophized another, "and if you ask me, we're all out of the same mold." He nodded wisely, pleased with his thoughts. "Men and toads," he said again, and went on down the bank and threw a stone in the water.

I put the sack in my shirt to protect the toad from interested violence, and started to wait out the hours. Over on the skyline the towers changed in the light. We would be here, I thought wearily, when the city had fallen, gross and neanderthaloid, sitting among our hatreds and superstitions, lighting our little fires in the gathering dark. We would throw stones and break what we could not understand, as before. It was part of us, that restless, manual cruelty from some dark tree in a vanished forest. It was our glory and it was, at the same time, our ending. I felt, though young, the long shadow of the coming night, and a great loneliness gathered in me. I spoke to the toad in my shirt, but of course he made no answer. He could only puff and tilt in a fatuous friendliness, not realizing there was no more desert and that he was engaged in a dangerous journey among monsters.

Perhaps that was true of ourselves, I thought finally, drowsing there by the switch lights. There was an outside world this toad was not equipped for—and men were toads—what had the old man said? Men were like toads in a desert of their own. Perhaps we were too limited to understand it. Like the toad in my shirt we were in the hands of God, but we could not feel him; he was beyond us, totally and terribly beyond our limited senses.

The rails began to creak to a far-off premonitory pressure and the little band of men started to string themselves along the grade with running space between each. The approaching locomotive headlight bathed everything in a calcined glare. I stopped a minute by the old man, shrinking back as the light went by and the dark cars began to roll past.

"S'long kid," he said. "Where'd you say you were headin'?"

I ran then and cried over my shoulder, "East to college," with a sudden impulse of daring. No one heard. No one was intended to hear. It was only a thought, an idle boast by now. To hell with it. Belatedly I joined the chorus. The iron ladder came by and I grabbed it and swung. My shout was lost in the roar of the wheels. I climbed the ladder and groped for an open trap door as the engineer began to open up ahead. The rocking roofs were briefly lit by the open firebox, and there was a great, passing glow on the night.

On the car roof I crouched a moment. Then I went down into the empty ice compartment and the toad and I were alone in the lunging, hurrying blackness. I let him go in the morning, in some sand along the right-of-way. It was the best I could do for him, and I hope he made out. His trip was ended, but I had a longer one to make, though in the end I suppose I saw no more than any tramp may see by the mechanical swinging light that greets one at lonely crossings.

I was nearly home but I hesitated. Maybe that midnight ride down the plains had hurried me too much. Maybe I wanted time to think. Maybe the life of the road was slowly winning me over as it did many men in that time so long ago. Maybe I was just perverse and, nearing the town of my birth, wished, because of old and bitter memories of childhood, to swerve aside.

Suddenly from speed I turned to lethargy. I dropped the train. I dawdled about. I slept in an abandoned freight car on a siding and made friends with a stray mongrel I knew I would be forced to abandon. One never should do this, but I found him a little food and shared it for a day or so. I still feel the pain.

Thoughts of home, school—little by little they were drifting away again. Without will, without intent, I was wandering slowly westward again into the wheat. I had reversed directions.

I was drowsing on the side runway of a Santa Fe oil car in some little place called Lawrence. I was back in Kansas, that was all I knew. The sun was warm. We had passed a river at intervals, a feeder to the Missouri. It was really a historic river, the Kaw, but I knew nothing about that. I was thinking I needed a bath. We stopped there in Lawrence while they dropped a car or so.

Suddenly I was stung awake by a cinder, luckily not a heavy one, but it struck my face. A volley accompanied by curses followed. I should have known. The ever-efficient Santa Fe. God, how I hated that road. If it wasn't their own security police, they knew how to encourage the town cops.

I dodged another volley by getting to the other side of the tanker. The train wasn't moving and the cop was. I vanished in the weeds and moved forward in order not to lose the train when it pulled out. A small town like that wasn't apt to make a big deal out of one sunning hobo. I was right. The volleys ceased, the cursing died, and somewhere far ahead the train be-

gan to roll again. That was a historic moment, although I didn't
realize it at the time. Some six or seven years later I would begin my teaching career in that same town. I didn't even know
then that the state university was located there.

It may be thought that the rocks and flung cinders were beginning to have some salutary effect upon my education. Quite
the contrary. I never entertained in after years any affection for
the local police force. They reminded me too much of Dodge
City in its heyday. If they represented educational stimulus I
would still be traveling. I never went until much, much later
up the hill to Mount Oread, where the university was situated.

Indifferently I floated farther into the harvest lands. Somewhere I left the Santa Fe and moved north into Rock Island
territory. I wandered into the Flint Hills. I had had my bath in
the Kaw. The town of Lawrence was just one of many towns
forgotten. I suppose men I was later to know were teaching on
that hill, but I didn't know there was a hill, or an Athens in
Kansas. I continued my meanderings on foot. I wasn't tearing
through the night on express trains any longer. I avoided cops.
I avoided jails. By no one but the law could I be regarded as
dangerous. I was just floating, waiting for something I didn't
understand. I let trains pass me, crowded with men California-
bound. I had a friend there who claimed to know where a little
gold might be panned. I never followed it up.

Still, I dawdled. I moved, yes; one had to move to live. I hit
the little bakeries. I lived, but in a wilderness of slow freights
and sunflowers. Sometime, I knew, winter would come. In the
meantime I was content to bob about in the shallows. If there is
any truth about these deceptive shallows, which I doubt, I was
finally among them. I was as lost as the mongrel pup I had been
forced to abandon.

If anyone taught me anything about love, it was that dog. It
is almost fifty years since I last saw him running desperately
beside the freight to which I clung. I didn't even have a name
for him. I wish we might meet somewhere. I hope, like my tilt-

ing, fatuous toad, that he survived. But I know better. I am almost seventy. I have lived a rough life. I know that neither of those two made out. I know also that I will never see that dog again. I may have given him his last meal.

Let men beat men, if they will, but why do they have to beat and starve small things? Why?—why? I will never forget that dog's eyes, nor the eyes of every starved mongrel I have fed from Curacao to Cuernavaca. Nor the drowning one I once fished out of an irrigation ditch in California, only to see him limp away with his ribs showing as mine once showed in that cabin long ago in Manitou. This is why I am a wanderer forever in the streets of men, a wanderer in mind, and, in these matters, a creature of desperate impulse.

CHAPTER 7

The Most Perfect Day in the World

I T was the last of my drifting days and if anyone were to ask what year it was—what month, what afternoon—I could not answer. I would be able to say only that it was for me the most perfect day in the world and that is why I retain its memory, safely severed from time and reality. Every man must treasure such a day into which he can retreat when the years grow desperate. It is never the same for each. For some it will be the memory of a woman, or a fading bar of music, or a successful night at a gambling table leaving you with the momentary illusion that you have won the game of life.

Also the perfect day is apt to be so subjective that no one else who was with you will remember it in the same fashion, if he remembers it at all. It will be a day totally yours. That is the way I shall think of my own day. I cannot name my companions, if indeed I ever knew their names. I only remember that there were four of us. But out of all the towns and stations of those years, it was somewhere in Kansas in the wheat. Was it Norton, was it—no, I think it must have been Phillipsburg. How we gathered there like the flocking autumn birds we were, I do not know. It was a chance meeting by a water tower and a loading platform, in an utterly wasted day.

The town was small enough not to bother us, and out of some

trifle of change we bought grape pop made by a factory in that very town. We drank it slowly with gusto and accounted it great, while we stretched in the shade beneath the water tank or lay dozing on the rough planks of the loading platform. I do not know whether we were headed west or east, or from what train we had dropped, or for what we waited. We were just there. Birds of passage with no past, no future, no desires.

We stretched out in the perfection of youth and health, grimy with engine smoke, blackened by the suns of a thousand miles. We laughed and took our ease and the world could wait for us. The world can always wait when one is young. It was not etiquette in those years to ask where a man came from or where he was going. Mostly he was going nowhere, no matter how far or how fast he traveled. What rode in his mind was his alone; he might be a thief, a gunman, or simply a man down on his luck. In the next forty-eight hours he might fall under a train and die, or over the horizon the law might be waiting for that same luck to run out.

No, there was no reason for any of us to hurry, but really that was not the point. On the rough boards where we talked and drowsed it was safe. We sensed subconsciously, I think, that we were out of time, secret, hidden. It was early autumn and the heat not oppressive. We drank the pop, bottle after bottle, like ambrosia, like forgetfulness.

I had been riding the tender of a locomotive the night before —that, at least, by straining, I can recall—and the youth with me had come creeping over the roof of the fast passenger to join me. When I had seen him starting to clamber down I had sweated uneasily with fear. I was too big for that kind of acrobatics on a hurtling train. He was the only man I had ever seen with the grit and the agility literally to dance with death upstairs on a flying express. He was one of the most perfectly coordinated men I had ever known and I suspected was a fine lightweight fighter.

Then there was an Indian—Mexican perhaps would be a

better word—but the thing was, he looked like someone who might have ridden with Geronimo. An utterly wild face that by some genetic twist had floated down from ice-age times. He could have been one of Attila's men, or equally have drifted with the first hunters over the land bridge to America. Every flash of expression had an animal's intensity. He studied faces like a trapped beast about to lunge. Perhaps it was a way of surmounting the linguistic barrier. *"Esta bueno,"* I groped, passing the pop when I knew it should have been tequila. I patted my stomach. *"Bueno,"* I emphasized. "A ya," he nodded, showing white teeth for a moment. "A ya," he repeated gratefully and drained the bottle. I had the feeling that if we had decided to assault the local bank barehanded, he would have come along happily. Wherever he had appeared from, it was another time.

That was another thing about the road. People were always appearing from some other century, entering and exiting, as it were, at will. You never knew whether your companions were from the past or the future. Since by common consent we had no real existence, we might as well have been teleported from the future as the past. But about that it was certain that no one would talk either.

The fourth man could have lived in any century and survived there. He was slight and brown and aquiline-nosed—a true aristocrat of the sort incised on Egyptian monuments. He should have been carrying a scroll of papyrus, but like the nimble-footed scurrier along roofs at midnight, he wore a pair of goggles high on his soot-streaked forehead. There were no Diesels then. If you rode the flyers you couldn't risk a cinder spark in your eye on a hurtling roof. It was touch and go at best.

As for me, I dwelt nowhere but in the unformed malleable present. Someone once said one should invent one's destiny, but if so I was devoid of inspiration. I merely waited and observed, having none of the skills these others had acquired. I waited and admired them all. I possessed, perhaps embodied, the shiftlessness of the times. Perhaps I knew it, perhaps I didn't care. I was introspective enough to welcome these men of differences, talents not my own, ethics of ages past or oncoming. I was merely lost, waiting to find a role for myself. Other youths, in the world I had left, had fathers who pointed the way. I had none. All I had was knowledge that the world was complex and dangerous. I was young. Someday something would happen to decide my course. Meanwhile I lay in the sun in Kansas and drowsed.

There was the boxer from Olympus who should have been crowned with laurel instead of goggles. What message had he for my future? Nothing that I could determine. He was somewhere back in the pure sunshine of Greece, where time had stayed its hand. Even today I cannot think of him as old. One's mind simply rejects the thought. No, he had come like a good comrade from beyond the centuries to lie with us on the splintered platform as though on the ship of Odysseus. I leaned on one elbow and watched him sleep, one careless arm flung out in a way that only the utterly nerveless and coordinated know— resting the way a steel spring may be said to rest. He brought no message, he never spoke. He merely existed at such a peak of physical intensity that everyone around him, including myself, seemed marred, imperfect. That was why I easily guessed that his was the century of the Golden Age. But one never asked in such company; that was why the centuries rested quietly beside each other.

I propped myself against the loading shed and spoke hesitantly to the Indian, trying another point in time, but not in a way to break the spell. He leaned forward out of a dark millennium, fierce, wild, intent, studying my face. "I have been in your country," I said, dredging up a few bits of Spanish and gesturing with a wide sweep far to the south.

"A ya," he said again, his teeth flashing.

"The Sierra, Sierra Occidental," I pronounced, *"muy alto,*

mucho alto, frigido." I clasped my hands across my body and rocked forward.

"Si?" he questioned at last, and I named a village far away in which blanket-shrouded men came out at night and sat immobile by a tarred road which had no traffic and no meaning. The Indians crouched there stoically to experience the heat still radiating from the vanished day. Once I had seen a face like his among them. They huddled and absorbed the heat and later in the night, while my teeth had chattered, I had flung myself desperately full length in the road to devour its dying warmth.

"Muchas estrellas," I said, *"y frio.* Stars and cold. High and cold, *comprende, señor?"*

"Si," he said, more kindly, but he raised his eyes to mine. Somewhere behind his pupils glimmered the backbone of the Americas since the ice. The last mammoth were there and the long cold down which this man had traveled, the highlands of two giant continents.

"A ya," he repeated, but this time he was far away. *"Las montañas, los llanos,"* he muttered with the intonation of that other tongue which had yet the accent of another language hidden beneath it. And another below that, I thought, till one gets down to naked ice and fire and meat. It is still there in his face. He has come a long way from the uttermost cold, but now he cannot find his way and watches me like a wolf seeking direction.

"Esta es la casa sin tiempo, nada," I said to his uncomprehending face and extended him another bottle while we shifted place a little to be in the sun.

The man with the face of one who might have held a scroll in Egypt spoke. "Ever heard of Atlantis?" he said.

"In books," I said, astonished, turning to face him.

"The books are wrong," he said. "The location is wrong. They knew better in Egypt but all that is gone, too. The time is gone."

Years and years later, I learned archaeologically what he seemed already to know, but would only speak of cryptically. "It was a beautiful place. The palaces of the Seven Kings."

"I wouldn't know," I said.

He pushed the goggles higher on his forehead and closed his eyes. Finally we all slept, the centuries together, pop bottles piled at our sides. I was the only unformed thing existing there, and perhaps it was intended for me, this dissection of time. I had an ancestor who had fought, so family tradition had it, with the Lion Heart at Acre. I wished he, too, had come.

We slept without concern, with no ostensible future there in the late summer sun. We had no destination. It was the perfect day, the centuries had finished their work, while we dreamed there in the shade of the water tower. There was this utter pause. No trains ran, no officials passed. Even the dogs slept in the village street. I turned over, I was the only one who did. It was the most perfect day in the world, the day time stopped. And I knew it, you see, I was young, but I knew it even then. That was the miracle, that is why I have remembered this one day. The day time stopped out of all my days. And the boards of the loading platform—that is an illusion. They were the planks known to Odysseus and that sleeping confidant of death who never said anything at all because nothing needed to be said. The day time stopped at a single water tank. Just that one time, no other, and to what centuries did my companions return? Only belatedly I speak of an incident more recent, for I, like them, am also receding.

Far up on the Wyoming border long ago I once came upon evidences of the last contact between wild, free-living Indians and the people who were to replace them. Scattered in sand blowouts by the Union Pacific right of way, hidden in arroyos running out of mountain uplifts, were exposed chipped splinters which were not the flints of an earlier period, but instead were the products of stone-age techniques exerted upon the discards of white civilization. Trade hoes had been ground into iron arrowheads now slowly rusting away in disuse. Lying about in the sand were what at first glance appeared to be stones but on closer examination proved to be fragments of opaque glass containers of late-nineteenth-century salves and potions which had still been current in my childhood. Here, scavenged from military dumps, these broken remnants of the white man's world had been reworked by pressure flaking back into the hide scrapers and knives of the last bison hunters. Here under the timeless High Plains sunlight, the primitives had tried to re-shape the new materials of another age than their own into forms they could comprehend. One could visualize, with only a slight effort, an impoverished dying band, crouched in a coulee turning their few gleanings over and over, trying hard to think how they might be readapted into the practical ways of the hunters.

Today in my age, in the great technological centers that have sprung up since my boyhood, I can understand that effort which even then I had gazed upon in a kind of fearful nostalgia as though I had felt the oncoming wind of change upon my back. Now I fumble with containers both of plastic and metal which, like puzzles, I can just barely learn to open, and toward which I am beginning to feel faintly inimical. Computers collect data about me which proves on occasion to be incorrect and which looses upon me the undeserved wrath of government officials. Big business similarly launches upon me remorseless demands which cannot be combated because the computer ignores my letters of protest.

"Pay up or go to jail," is the sole response of the heartless machines. "Pay up or your credit rating will be ruined." Fumblingly I examine these computerized responses like a primitive seeking to understand the civilization with which I once had some personal contact. Unlike the nomads crouched in the lee of the railroad embankment, however, I can find no way of reshaping these faceless notices into an arrowpoint. Within my own lifetime I have completed the journey from my perfect day as a wanderer to an old man crouching by a wall in the October sun, glad to be left alone to his observations of insects and such small mammals as remain. Already I am as much an anachronism as the last warriors lingering helplessly along the railroad line that had destroyed the buffalo and the Indians' way of life.

The structure will fall, I prognosticate, but not in my time. Still, I salvage a thick chunk of workable glass from a fallen streetlight. I also bring home a huge bolt.

"What is this?" asks my wife suspiciously.

"It will make an excellent club, woman," I growl, the warrior in me stirring. "Now that they do not permit us to have guns, this may be important." I put the bolt upon a shelf and think, comfortably, I fear, of what happened to Custer.

It is all in vain. Next day the bolt has disappeared. I am allowed the glass to practice pressure flaking. The huge man-killing Paritintin bow I once had has been given away to a museum. I brood upon it darkly and, using my scholar's disguise, draw out a book from the library upon crossbows.

It is all fantasy, of course; I will not live to witness the fall or the return to these things. Still, the hand ax in my drawer has more meaning to me than an antique. Now and then I draw it out surreptitiously and make sure that it fits my palm. I turn it over meditatively and murmur to myself when no one is looking. I must be careful. My maternal grandfather spent his last years in considerable isolation pawing through a great sea-chest of tools, like a mad Viking.

I must be careful, I think again, feeling covertly the archaic weaponry in my desk. I must not display such interest in a club-like bolt as I did in the case of that one which I had picked up in the street on the way to a dinner party. The bolt had aroused my wife's severe attention because she had seen it sagging in the coat of my dinner suit. No, I must sit in the sun by the back-yard wall now, just as those few last warriors by the railroad, converting trader's iron into arrow points.

"A ya," I start to say to one gone, "the mammoth will be passing shortly." My voice has also changed. I know I am speaking in gutturals long gone by—the gutturals of the mammoth hunters.

PART TWO

DAYS OF A THINKER

One must somehow find a way of loving the world
without trusting it; somehow one must love the
world without being worldly.

—*G. K. Chesterton*

CHAPTER 8

The Laughing Puppet

I HAD never known anything to last so long. Eight years and in them death, illness, boredom, uncertainty, at a time when others went straight to their careers, or to whom doors were opened. Eight years that might as well have been a prison. Perhaps they were, in that I could not get outside the ring, the ring of poverty. Like a wolf on an invisible chain I padded endlessly around and around the shut doors of knowledge. I learned, but not enough. I ran restlessly from one scent to another. Sometimes I gave up and disappeared into the dark underworld of wandering men. Or I worked at menial tasks and convinced myself I would have been content if one of them had lasted.

I knew I was ill with influenza from the moment I awoke in the rooming house where I was quartered, some blocks from the university. It was the final day of examinations—and this one examination in particular, if passed successfully, would complete the requirements for my undergraduate degree. Nominally then I would be labeled as of the class of '33, though the people I had started college with in 1925 were all vanished into the outer world, people with children, jobs, careers—doctors, lawyers, teachers.

I had flunked courses before in my interminable and intermittent despair. I sat on my bed amidst a growing feeling of sickness. What I thought was not pleasant. The head of that particular department in which my ordeal was to take place was a harridan who would have been creditable only as a guard at Buchenwald. If I appeared later I would have been sure to get the kind of examination she reserved especially for people like me.

I had acquired in those years some small fragments of worldly wisdom. The time was now. One more impediment and I knew what would happen. I would drift out into the world of violence forever. There was a letter lying on my desk from an old companion suggesting a prospecting trip in the Canadian Rockies. I dressed slowly, walked to school, and took the examination. My immediate instructor, a very lovely and understanding woman, was killed shortly thereafter in an auto accident. I had passed, however, and my record was never again to trouble the files of my university.

On that afternoon, without knowing the result of my efforts, I slowly climbed the second-floor steps of my rooming house. As I fumbled for the key to my door I suddenly heard, very far away, as though in another part of the house, the small faint thump of something hitting the floor. That was the last I knew until I awoke upon the landing. Everything had grown dark and I crawled frantically about on my knees, dimly realizing that I had lost all sense of direction. Instinctively I seemed to know that there was a stairway back of me down which I might tumble. My location slowly came back through the sense of touch. I threw open the door and tumbled in upon my bed. Then the nausea struck me. Gasping heavily, I staggered down the hall to the bathroom and vomited. Again I reeled back to my room.

Though there were normally other people on that floor, no one had passed over the landing, no one had discovered my unconscious body. With the secretiveness of a wounded animal I lay hidden all night in my room. It was the sort of behavior, I suppose, that had become habitual with me.

THE LAUGHING PUPPET

Some of the black marks on my college record were products of the same suspicious fears shown by my deafened mother. Other students stumbling into the wrong course or encountering inimical instructors went to their advisors and legally dropped the subject. I simply walked away and there the record stands. Bureaucracy intimidated me. I had come from the world of the night. Once when I had rebelliously dropped out of Lincoln High School and found a job, I had difficulty eluding the truant officer. I was not criminally inclined, though I might have become so in a city ghetto environment.

I merely wanted to be left alone, but still I felt this persistent urge toward books and toward those words of my father which I had seen crumbling in the flames, never really to be effaced. I took them as my only heritage. I tried to make whatever dream father had had of me into a reality. I found in those eight years that my appetite for wide areas of learning was insatiable, but there was no one to guide me. There was no one to say, "Be a doctor, be a lawyer, be a teacher, a historian, a writer." Perhaps I was none of those.

Once, in high school, I had written, more or less blindly, an essay for an English teacher. "I want to be a nature writer," I had set down solemnly. It was at a time when I had read a great many of Charles G. D. Roberts' nature stories and those of Ernest Thompson Seton. I had also absorbed the evolutionary ideas of the early century through Jack London's *Before Adam* and Stanley Waterloo's *Story of Ab*. None of this had come from high school. It had come from the books brought home from the local Carnegie library to which I used to pedal in my coaster wagon. "I want to be a nature writer." How strangely that half-prophetic statement echoes in my brain today. It was like all my wishes. There was no one to get me started on the road. I read books below my age, I read books well beyond my age and puzzled over them. In the end I forgot the half-formed wish expressed in my theme.

I had to seek food, shelter, and clothing. I remember a philosophy professor for whom I once read and graded papers—a "research fellow," they would now call me in this more grandiose time. One day I saw him eyeing the ragged shoes I was wearing. "Perhaps you would like to proofread some chapters of my book," he ventured hesitantly. He had come from abroad and English was a second tongue to him.

"I would be glad to help," I said. Later I found several of these useful tasks among other professors. The situation contained its ironies. The first English course I had ever taken at the university had turned me away from any thought of a formal career in that subject. The teacher had read my first assignment and told me bluntly, "You didn't compose this; it is too well written." Good or bad, it happened to be my own. The man couldn't prove his own assertion, but again, with the wary withdrawal of an animal, I merely turned away.

Today I have many friends and acquaintances in the world of English letters. In college I took little beyond the minimum requirements. There were professionals who welcomed me to their offices, some who were kind, though with whom I had no contact in a course. The point is that this early unwarranted accusation had destroyed the pleasure of the formal subject for me. I was far on into middle age before I found a belated joy in professional literary studies. Perhaps that one belligerent sentence had something to do with my turning aside into science.

Yet looking back I am not convinced. For a thin-skinned young man, still emerging from long isolation, encounters with the realities of the academic world were not always pleasant. My feelings for the western lands, uplifted scarps and buttes and the things contained in them, had long outweighed the written word in the emotional content of my mind. It was, perhaps, my mother's stifled vision, the visual impact of places demanding to be celebrated in essays I did not know how to write.

But what if I had not recovered consciousness upon that upstairs landing? When one faints there is normally a little forthcoming awareness, a final chance to get down before one falls.

THE LAUGHING PUPPET

I had dropped as though a bullet had felled me. The thud of my head hitting the floor had sounded so remote I never had time to relate it to myself. I simply ceased to be. There was a body lying in the dusk on the stairs, but whatever motivated that body, consciously dwelt upon its problems, looked after its needs, had vanished. Yet somewhere in that silence where I no longer existed, the heart had picked up and labored in the midst of a retching sickness. Swimming and clambering phagocytes had swept through an arterial system that had become a battleground, a Gettysburg of charging forces. John Hunter, the surgeon, once spoke of a principle of incompleteness. I suspect he was groping toward whatever it may be, tangible or intangible, that sews up our ragged heads and limbs if the stuffing has not all run out of them.

The experts say that within six minutes of heart failure a man's brain, which consumes enormous quantities of oxygen, will be injured beyond recall. That with prompt medical attendance and mouth-to-mouth resuscitation one may be brought back uncertainly, perhaps, over a thirty-minute threshold. What is shrouded in mystery, however, is the way by which the sprawled body, even if the heart has not ceased to beat, reassembles itself as an entity and relights the flickering candle of consciousness.

I was gone so far into the dark that no dream whispered to me, no sound again troubled my ear. Consciously, I did not exist. But something was still alert in me, the Synthesizer. A hurrying, jostling cavalcade of hemocytes was rushing oxygen toward those cranial abysses into which I had vanished. Something, some toiling cellular entity of which I was unaware, was searching me out, reconstructing me, setting failing ganglions to sputtering, reactivating all manner of wildly spinning compasses. Today as I read of the achievements of physics, of those leaping, impinging, flying phantoms out of which is created the nature that we know, I cannot but think of ourselves as some microcosmic parallel.

DAYS OF A THINKER

Our blood contains ingredients as mysterious and as abounding as the micro-particles below the atom. Our spider web of neurons holds in its tenuous interwoven film our memories and our conscious existence. Yet where do those memories abide when the dark comes down? Who seeks to restore the snuffed-out candle? Why should there be this enormous effort to reactivate a sleeping manikin? From whence comes the organizer, sending millions of cells to do battle in the dark and from the body of this death to haul up one solitary Lazarus?

Oh, I know the Synthesizer has his limits, but still he sent my father back just one more time a long and weary distance to my brother. In all my remaining years I have been grateful to those unseen toilers who, when my will had failed, had re-created what individually they neither knew nor cared about.

A youth so young who has felt all sentience depart, and then has experienced in his own body the strange climb up the crags of darkness to the lighted tower once more, will not, I think, be satisfied that the universe can be found out. It hides its work. Look close and it dissolves in aimless atomies. Look again and it is slyly engaged in performing miracles in a web of its own light. Since that disastrous moment on the stairway I have never been sure what nature is about. Sometimes I see it peering at me from the pigment spots on a snail's horns, or, many years later, I once heard it, in the shape of a cat, address me upon ethics.

From a single experience I had learned I was a genuine stranger in the cosmos. No blood cell, no single neuron would ever inform me how the light of consciousness had been relit; they had made of me a construct. I was the lonely one in whom they swarmed in their millions. I was their creature; alone they had re-created memory and light. Once more I, the construct, am eternally grateful.

Does this sound as though I were descending to maudlin sentiment? Scarcely that. I am merely recognizing the existence of mystery, of the unknown. In fact, after my silent recovery from the episode on the landing I had, in the beginning, small reason to rejoice. Some sort of interior thermostat had finally broken. I could not sleep when I wanted.

Perhaps insomnia had started after my father's untimely death; perhaps it had been augmented by the fire-ridden anxiety of those nights at the hatchery which, ironically, I learned had been consumed, years later, by just such a fire as I had anticipated. When I did sleep, I dreamed excessively. Something about these dreams led me to wonder about my childhood as another part of my sleep syndrome. As I have intimated, I was a solitary child in a divided household. Far earlier than is often recognized by parents, children come to the realization of their own potential deaths.

For Robert Louis Stevenson this knowledge was involved with the lych-gate of the Warriston cemetery, "a formidable but beloved spot." For William Golding, the novelist, it was the dark cellar of his early home which lay dreadfully close to the cemetery, so close that the dead, he thought, "must have their heads under our wall." As for me, these childish terrors involved the darkness under the bed in an old house where I had once by lamplight studied a gaudy picture in a newspaper.

The picture had shown an array of demons with pitchforks prodding a host of souls into a flamelit hell's mouth consisting of the jaws of some giant reptile. My own childish preoccupation with terror led me to play with my blocks dangerously near to the dark shadows beneath the bed. Sometimes I fancied there was an invisible presence reaching for me if I came too close.

Children live in a Lilliputian world. I was closer to that mysterious realm of darkness under beds and sofas than were my parents whose minds floated high above me in the light of lamps. We lived in a kind of Victorian dinginess. The houses were gloomier than today's houses; the kerosene lamps were dimmer than the lights of today. As a small boy I lived the farthest down in that shadowland. I was the closest to the demons and succubi depicted in Sunday supplements. Adults, it appeared, knew nothing about death. It was the very young who were forced to examine it.

But about the mechanics of all this I have no sure knowledge. I only know sleep no longer came at will. My father returned once in the years that followed—in a dream, that is. He was diffident, an air of trouble hovering about him, a sense of not staying. I felt it in my mind, but neither of us said anything. He was growing tenuous, disembodied. I saw him once more through the dark pane of a window glass in the night, then he was gone forever, to be replaced by something far more terrible than the gentle dead—death itself the secret agent dogging every alleyway of my dreams.

Sometimes I pushed furiously against a door and something invisible, sending crepitations down my spine, pushed back. Once, alone, as in the hatchery days, I walked a house at midnight, going round and round a great oval furnace, sensing something pacing just ahead of me but always out of sight. Sometimes I struck out in the dark at nothing; once I shattered a lamp, the crash awaking me. Nothing was ever seen, there was only the ebb and flow of this formidable force, this creature which I could neither retreat from nor successfully confront. Through years it continued, a silent unseen duel.

Just once, I caught him, or, rather, I almost did, not long after my uncle's death. I lay exhausted in my room which slowly in a dream became the parlor of my uncle's home. I was there sitting in a remembered rocker, waiting. There was a low laugh behind the curtain of a door, the sound of a snapped lock. The laugh continued, deep, vibrating. The lights went out. There was a voice. "We are alone now. Isn't that what you have always wanted?" The voice came from my uncle's chair. Without hesitation I hurled myself upon the chair. An equal force arose to meet me. We fought until, across the wreckage of the room, I finally realized I was triumphing. I had something I was shaking by the throat at last, something collapsing beneath my hands. Inexplicably the lights once more came on.

I grasped an unrecognizable, collapsed puppet, a thing of *papier maché*. I had crumpled its throat in the red rage of despairing murder. It lay weirdly distorted upon the mahogany sofa. I stood back, panting. It was a mask, a mask for the escaped invisible intruder. Crushed laughter still wrinkled its features. I sat down and thought weakly, clearly, for the first time: perhaps that's all there is, an emptiness. He meant to show me—

I lay on my back amidst the rumpled covers of the real room while that other slowly faded, the chair, the shattered, laughing puppet, all that held me to my past. I was a graduate student at the University of Pennsylvania now. I held a scholarship. The past, it seemed, lay dead as the broken puppet. Eight years. I had fought my way outside the ring of circumstance at last. Nothing would ever bring me back. Nothing. I was, it seems to me now, still very young.

CHAPTER 9

The Badlands and the School

I WILL never forget my first day of registration at the University of Pennsylvania. I had come directly from the *Mauvaises Terres,* the Tertiary badlands of western Nebraska, into a great city of banging, jangling trolleys, out of a silence as dreadful as that of the moon. As a fossil collector for the Nebraska State Museum, bones, not people, had been my primary concern. Field parties in those days were not equipped with portable radios and television sets.

Save for occasional visits to far-off towns, no news penetrated to us. We lived in a timeless solitude that had already existed when Egypt was first rising from the mudflats of the Nile. We wandered among gullies and pinnacles cut by wind and rare desert rains. Vegetation stopped at the edge of those declivities; there were canyons, volcanic clays over which toppled sandstones, tilted like the giant menhirs and dolmens of megalithic Europe.

Few people outside of the realm of paleontology realize that these runneled, sun-baked ridges which extend far into South Dakota are one of the great fossil beds of the North American Age of Mammals. Bones lay in the washes or projected from cliffs. Titanotheres, dirk-tooth cats, oreodonts, to mention but a few, had left their bones in these sterile clays. The place was as haunted as the Valley of the Kings, but by great beasts who had ruled the planet when man was only a wispy experiment in the highlands of Kenya. These creatures had never had the misfortune to look upon a human face. Most of what we knew of mammalian evolution in North America had come from this region. All the great paleontologists had worked here. New species of animals still occasionally turned up.

The place enchanted me. I have an almost eidetic recall for those solitary years. I owe my presence there to C. Bertrand Schultz, a fellow student in geology and anthropology, who later became field director and still later director of the Nebraska State Museum. He was one of those fortunate people who knew his course and did not wander. Today we are both approaching the shadows of retirement; he is one of the leading authorities on the successive faunas and geology of the middle border. I am profoundly grateful to both him and his predecessor, Dr. Erwin Barbour. Much of what seared its way into my brain and into my writing came about because Schultz prevailed upon the museum director to allow me to work with the field party. In the timeless land, I could remain hidden.

All things end, however. With some small savings from a bone hunter's salary, plus a certain faith and assistance on the part of my uncle, who had been swept into state office in the Roosevelt landslide of 1932, I found myself before College Hall on the University of Pennsylvania campus in Philadelphia. In an ever-increasing racket, I made my way from one office to another, receiving approvals and obtaining signatures on cards. Suddenly the noise, the cacophony of horns, became nerve-shattering. After all, I had spent a long summer in the silences of vanished geological eras. The urban world was, for the moment, unendurable to me. I went up Woodland Avenue and came to the gates of a deserted cemetery. With the sure instinct that time would vanish here, I walked far back among ancient mausoleums and more humble monuments. Seated at the foot of a giant obelisk marking the remains of a forgotten general, I sat until the sun waned toward evening. Today I would give myself no more than a fair chance to emerge from that cemetery alive. At the time it was safe enough. Finally I wandered out, reading the names of the forgotten as I passed.

In a few days my tolerance of noise levels increased. I went to a nearby bank to deposit my small savings. "Sorry," said the clerk grandly, tossing the money back. "We do not accept deposits of less than five hundred dollars." Shortly after, the deepening depression caught up with that bank. It vanished. Banking, it seems to me, has become in Philadelphia a much less arrogant business than in my days as a student. At that time it appeared coldly oriented toward the great accounts. The building where I had my first rejection still stands, but it is now devoted to medical offices. Perhaps the clerk's unpleasant condescension served me well, as it turned out. I still wonder briefly, as I pass the building, what breadline he entered.

I climbed then, slowly, up the four flights of stairs to Dr. Frank Speck's office, under the old tower that has since vanished. A pleasant, attractive Indian girl occupied the desk by the door. I was fully enrolled now. Dr. Speck was seated at the little desk I was to come to know so well. His office was large, a former classroom, well lighted, high, and airy. All the wall space was lined with books and folders filled with off-prints. Speck, I was to discover later, disliked the central library. If he had to consult a file of journals, he preferred to send an emissary, which in my later graduate years turned out to be myself.

Mostly he labored in his own office amidst distractions which personally I have never been able to endure. Students lingered between seminars which he conducted usually at a huge table in the center of the room. He wrote letters or papers in a big bold hand at his desk while people murmured about him or plied "Doc" with questions. I find difficulty in describing him now, even though a faded informal winter photograph sits before me as I write. He is clothed in the garments of a trapper of the Canadian forests, part Indian, part of white manufacture. Snowshoes are gripped in one mittened hand. The picture was probably taken in Maine or Labrador in the prime of Frank's life.

Under the muffled fur across his forehead one can detect the outlines of a bold, formidable countenance that would have been acceptable among the mountain men who long preceded the rush of white settlement into the high plains. He was a stocky man, slightly below the average in height of this generation. The mouth in the photograph is firm, with a little cigar of the brand known as Between the Acts firmly gripped in a corner of his mouth. The head is massive; a total, lone-wolf independence shows in the stance of the body. The face is in no sense aquiline or Emersonian in the old New England sense. It is that of a belated sea captain, decisive, capable of the use of a belaying pin in an emergency. Or the face of one of those explorers trudging on relentlessly through the waterless seabed of the Great Basin. A man, in short, definitely not of the age he inhabited.

Used to another environment, I, and several other new students, departed promptly at the end of each class hour. "Doc," or "the old man," as we variously called him, appeared surly, or at the very least sulky. I had come to share confidences with another student, Ricky, whose father, then president of a Southern college, had previously taught in an American school in Tokyo. Ricky had grown up in Japan and could speak the language like a native. For an American this was an extraordinary achievement. We became fast friends but worried over our seeming inability to make much human contact with our departmental chairman. To make it worse, one day in class a little heap of square-cut flints had been poured upon the table—one of the surprises with which Speck amused himself.

"What are these?" he barked. "Any of you know?" He looked at me. I was supposed to be the archaeologist in the group.

I examined them, and a western memory with an absolute surety came back. "They are not Indian at all. They are eighteenth-century guns flints for flintlock rifles."

A titter ran about among the students at the table. They were sure the old man had tripped me.

"What makes you so sure?" he said menacingly, trying to make me back down. "Sir," I said, "I just know. They're square-cut European flint. I've seen them on the guns themselves. I am sure I'm not mistaken."

"You are right," he growled reluctantly. The tittering ceased. "Class is dismissed."

"Jeez," whispered my companion. "The old man will make you pay for that. He likes to win those games. Why didn't you shut up?"

"I couldn't," I said logically. "He asked me directly. I had to answer." Ricky shook his head.

A few days later, a dark, broad-shouldered man a bit older than Ricky and myself hailed us on the street. He had not been a member of that class at the table but had been browsing in Frank's library. I had noted before that he had seemed a particular favorite. Rumors had reached me that he had been a noted athlete who still occasionally refereed games. Now he was an advanced graduate student in the department. He introduced himself as Lou Korn. I did not know it then, but he was destined to become a lifelong friend and stand up at my wedding.

"Look, you fellows," he said. "You're not getting anywhere with the old man. I'll tell you why, if you don't mind."

"Why?" we chorused. "We don't skip class. We study. What's the trouble?"

The athlete with the shoulders of a running back and a dark, ruggedly handsome face grinned quietly. "Because," he said, "you leave directly after class. The old man doesn't like it. He thinks if you're a true anthropologist, not just a student, you should stick around. It's part of his way of judging people. You," he poked me good-naturedly, "both pleased and put him out by solving that old gunflint trick of his. But he liked it.

Proved you're from the West, you know. Boy, are you going to get an education! Now, remember, when you see him on the campus, 'sing out,' that's his phrase for it."

"Honest?" we said. "You wouldn't kid us?" He shook his head. "Just remember the old man hates formalities and he's spent time in the north woods. Most of what you learn from him, you'll learn across the street at the Greek's. He generally goes over there for coffee and a dish of ice cream at three. His favorites," here he paused, "generally traipse along. Stick around a little more and read in the office. Wait till I give you the nod."

I learned a lot about Lou Korn in those few moments. Secure, himself a favorite, he had gone out of his way to inform two helplessly floundering newcomers who had not grown up as undergraduates under the old man's tutelage. Frank, in a strange way which I was never totally to understand, was lonely. His initial growl, which I had not anticipated from my correspondence before coming to Penn, was an assertion of his own independence. He expected perception. If you saw through the gruffness, if you liked books, snakes, the life of the hunting peoples, if you were a good companion, he would do anything to help you.

In return he wanted very little, an exclamation over a rare fern, something in the way of beliefs shared as though by two men who sat before a brush shelter in the flickering dark of a campfire—a fire and a dark that had not changed since man entered the world, a totemic dark in which animals spoke and skins were easily shifted.

Once, the Anthropological Association met in Philadelphia during my time there. Distinguished scholars, some from abroad, drifted into the office and were pleasantly received. But they came to Frank Speck. To my knowledge he never attended the meetings. "Loren," he said, "stay with me. Everybody's over there," he gestured. "There're more brains in Philadelphia this week than there is sewage in the Delaware River." His voice had lowered to a growl in which I detected the uncertain tone of a waif.

"Okay, Doc," I said, though I had intended to join the others. "Where shall we go, the pine barrens? The zoo?"

"I know a quick way over by train to the barrens," he answered. "We can tramp and canoe a bit and come back in the late afternoon."

The details of that day are long forgotten, but two men alone exchange confidences. His grandfather had been a sea captain. "These Pineys," Frank swept a hand over the woods through which we tramped, and I knew he was referring to the illiterate inbred woodsmen who inhabited the place, "were wreckers two generations ago. My grandfather's ship was lured onto the shoals by false lights during a storm. Grandfather's body was cast up along with the cargo and some of the crew. In those days that was the Atlantic, with its false lights, and the piratical people of the shore."

He paused and took a quick breath and went on. "My father had to drive down in a wagon and take the body home. It's odd, you know. I've sat around fires with Pineys whose grandfathers did the job. They would half admit it."

I thought silently of the brakeman long ago in the desert. I wondered if a wire flickered somewhere in the old man's brain. "My father," Speck continued, "became a broker in New York. I was sent to Columbia. I was intended for the ministry."

There was a gap in his history I was never to fill, except in fragments. Out of the reticence of long road experience I never asked questions. I only listened to what people were willing to volunteer. Though Frank's mother was still living when I knew him, there had been a time in childhood when, in ill health, he had been entrusted to the care of an old Mohegan woman. Why this was I never completely understood, save that this foster mother was a family friend. Had his real mother been ill? Or had his parents thought that this would be a valuable experience for the youth? I was never to know. One thing I did know, however: this episode indirectly brought about Speck's eventual meeting with Franz Boas, then the dean of American anthropologists.

In the first decade of this century American anthropologists of distinction could almost be counted on one's fingers, and the places where anthropology was taught as a separate subject were few. Many of these men had actually emerged from different disciplines, drawn, perhaps, by the wild and uncontained boundaries of the subject. Boas, himself the teacher of several anthropologists who later attained great eminence, was a maverick physicist who had written his doctoral dissertation on the color of seawater.

The sudden rise of the subject was phenomenal. The question of relevance was never raised. In fact, if relevance, as recently defined, had been used as a criterion, I sometimes wonder if the science of anthropology would have survived. Yet by some paradox it became remarkably popular in the fifties and sixties. Perhaps by then it had come to represent to the young the abolition of ancient taboos and the rationalization of their own life style. Furthermore, it had contributed to the elimination of much ingrained racial prejudice and to a better understanding of the movement of cultural traits around the globe. These are merely observations made in passing and must be so taken. As in the case of any science, not all its practitioners need be regarded as reasonable or without self-interest.

Of Frank Speck I know this much: that because of knowledge of a dying tongue, Mohegan, derived from his Indian foster mother, who taught it and much else to the impressionable boy, he was finally brought to the attention of Franz Boas and turned aside from the ministry into anthropology. The change in professions did not, I have come to believe, effect a total transformation of personality. I base this upon two observations.

Once, strolling in the Philadelphia Zoo, we came upon a wood duck paddling quietly in a little pond. These birds are most beautifully patterned. We stood watching the ducks. "Loren," Speck finally said, quite softly and uncertainly for him, "tell me honestly. Do you believe unaided natural selection produced that pattern? Do you believe it has that much significance to the bird's survival?"

I turned in surprise, because unbeknownst to my distinguished teacher, the same thought had been oppressing me. "I know," he said hastily, "what all my colleagues would say, but they are specialists on man. You have wandered to us out of another field. I'd like to hear what you think."

I tried to choose my words very carefully, not to satisfy a man or promote my own interests, but because, like Frank in his northern forests and amongst the wood people, I had been much alone. "Frank," I said, "I have always had a doubt every time I came out of a laboratory, even every time I have had occasion to look inside a dead human being on a slab. I don't doubt that duck was once something else, just as you and I have sprung from something older and more primitive.

"It isn't that which troubles me. It's the method, the way. Sometimes it seems very clear, and I satisfy myself in modern genetic terms. Then, as perhaps with your duck, something seems to go out of focus, as though we are trying too hard, trying, it would seem, to believe the unbelievable. I honestly don't know how to answer. I just look at things and others like them and end by mystifying myself. I can't answer in any other way. I guess I'm not a very good scientist; I'm not sufficiently proud, nor confident of my powers, nor of any human powers. Neither was Darwin, for that matter. Only his followers. There were times when Charles Darwin wobbled as we are wobbling here. Remember what he said once about the eye?"

"A cold shudder," quoted Speck promptly.

"Well," I added, as the duck paddled along slowly, displaying its intricately patterned feathers, "that's just the way I feel right now, as though the universe were too frighteningly queer to be understood by minds like ours. It's not a popular view.

One is supposed to flourish Occam's razor and reduce hypotheses about a complex world to human proportions. Certainly I try. Mostly I come out feeling that whatever else the universe may be, its so-called simplicity is a trick, perhaps like that bird out there. I know we have learned a lot, but the scope is too vast for us. Every now and then if we look behind us, everything has changed. It isn't precisely that nature tricks us. We trick ourselves with our own ingenuity. I don't believe in simplicity."

I subsided, feeling I had merely befogged everything with my confused ignorance. We stood in silence a little while and then went to a cage where Speck let a parrot gnaw his hat. Speck's hats always quickly took on the appearance of those I had seen on the road—even the one he had left hanging on my office wall before he had gone away to die. But that act lay years beyond us.

"Loren," Speck said as we made our way toward the gate. "I feel as you do, maybe even more, because I really live far back with the simpler peoples. You have seen a few come to the office. They trust me and bring things to trade."

"Yes, I know," I answered, "and neither of us can quite speak to our contemporaries. Certainly I don't. But I'll tell you something. A man named Algernon Blackwood, an English writer, once wrote a story about a man in Egypt whose soul was stolen by the past, by all those giant millennia heaped in a little space along the Nile. Oh, he remained here in the flesh, and walked upon his errands, but his mind had otherwise been taken. He had vanished into another age. Perhaps if we go on this way it will happen to both of us."

As a matter of fact it did. I was unknowingly prophetic and I was the last to go. For Frank, perhaps the process was either slower or had unconsciously been effected by his childhood Mohegan experience. Perhaps he was a true changeling, a substitute child in everything but appearance. The Indians of the Northeast, long supposed to have vanished or been ethnically absorbed, were always his primary interest. The girl I had met acting as his secretary was one. He had an uncanny way of locating remnants that were already supposed to have disappeared in Thoreau's time.

A later generation of ethnologists will have to rely in large part upon Frank's records. He found and recorded customs still extant that other workers had assumed were extinct. He lovingly gathered up the broken bits of Algonkian tongues from speakers upon whose lips they were dying. He recorded the last details of their material culture. He worked mostly alone, before the day of the big foundation grants.

It was just as well. What he was seeking was as elusive as a beautiful night moth. One found it by oneself or not at all. Students of primitive religion have always utilized his studies of the game lords, of scapulimancy as a circumpolar trait. Stuart Neitzel, an excavator as far down as the Bayou country, writes to me about him still.

I said earlier that I based my comments upon Speck's essential retreat to the primitive upon two observations. No doubt others could be added from the intervening years, but long after his death I happened to be speaking to Roy Nichols, Penn's nationally known historian and Pulitzer Prize winner. "Frank," he allowed, "was a pioneer in ethnohistory." Then, pausing, this cultivated scholar, in some ways the utter antithesis of Frank, ventured a comment upon Frank's difficult last days, his pride, his unwillingness to confide his distress even to lifelong colleagues, his pathetic subterfuges, his secretive disposal of his valuable library. Nichols closed the chapter. "Frank," he said, "was basically an Indian. He died one. Mentally he went back to the forest. You couldn't help. Don't blame yourself. He wanted no sympathy."

I sighed and thought of the wood duck paddling in its little pond. "Roy," I said, "we're all winking out, but Frank was something special. I left an Iroquois mask he had given me up in his office. I didn't think it would be good for me after he was gone."

"It's got to you then," Roy said. "You anthropologists go prying into tombs or forests and things come home with you. Be careful, be very careful."

"Yes, Roy," I said, thinking at last of the day in Kansas by the water tank and those who had drowsed beside me there. "By the way, did you ever hear of the palaces in Atlantis?"

"Not my specialty," said Roy, grinning, "but it's Crete, if you want to know. They tell me now Crete is Atlantis. Some mistake in Plato's figures."

"I was just asking," I said. "I heard it first from a man in Kansas."

"He was right," said Roy, "dead right." He strolled away. I lingered doubtfully a little longer upon the walk. It was best that I left the mask with the twisted grin, I thought obscurely. They were not the right people, the Iroquois. I belonged further back—back on the *altiplano* with the great grey beasts of the crossing. I would never meet Frank. I would never meet Roy. There was a far-off sound like the rattle of tiny dice in my head. I shook my bad ear. It was surely no more than that.

CHAPTER 10

The Crevice and the Eye

I HAVE told of the miniature gold crosses that were hidden in the weeds next to my home and that were stolen by a callous reaper. Perhaps they symbolize in a way the conflict at the root of my being. Always, standing above excavations, my own or others upon which I labored, I have been both excited about what the shovel would reveal and disconsolate and stricken at the sacrilege done to the dead. Once, high on the side of a sullen mountain bastion in Texas, our little scouting party had unearthed in a cavern a child's skeleton tenderly wrapped in a rabbit-skin blanket and laid on a little frame of sticks in the dry, insulating dust. An assemblage of bone needles and a "killed" rabbit stick broken to accompany the dead had been tucked away with the child. I stood silent and was not happy. Something told me that the child and its accouterments should have been left where the parents intended before they departed, left to the endless circling of the stars beyond the cavern mouth and the entering shaft of sun by day. This for all eternity. Yet I knew also the valley population was growing. Vandals and pothunters would inevitably discover the child's resting place. Eventually all would be crushed, broken, or sold for antiques in the valley below. In the end the devastation—on a smaller scale, for these were a simpler people—would be as great as that which had befallen the pharaohs. Here, perhaps, an institution might at least rescue what would otherwise be destroyed or dispersed.

"We can give all this to the local museum," commented the expedition leader abruptly. I turned in some astonishment. "It will help to cement relations, keep the locals interested," he continued practically. The one thing he did not mention was the fact that the local museum was largely a dream in the mind of one or two town collectors who had been instrumental in calling our attention to the shelters under the face of the cliff.

"But," I started to protest.

"Nothing," grunted the expedition leader, who was passionately addicted to the pursuit of terminal ice-age man. "We don't want to bother with this stuff. Let the locals have it. We've got to go deeper, much deeper."

I stood in the cave entrance a moment longer looking across the distant valley toward the Rio Grande, holding the burial cradle and the cuddled child in my arms. A man could do very little with his superiors and not be thought mad. I set the cradle gently down beyond the debris of our excavation. The picks went on falling to no great purpose in that place. What eventually became of the contemplated museum I never knew, nor of the child carefully wrapped in a rabbit-skin blanket and the tools intended for its after existence. Wasted, really, wasted because the man in charge had a driving mania for one thing alone—the people of the ice.

He will have his coveted place in the histories of archaeology, but, as for me, my mind ran painfully back to the gold crosses stolen by the wielder of the scythe in my childhood. I could have spent a day up there on the great range just listening to the wind and talking to the child, murmuring to it across the centuries. Now we would go down, and the cradle and its little occupant would be handed over to others. I have never revisited the spot.

Other men have had their mingled share of triumph and remorse in these episodes which have a way of striking hardest in great deserts. Crouching before Tutankhamen's burial chamber, deciphering with difficulty the royal identity seals, the scholar James Breasted felt the whispering, whimpering wind of change pass over him as the fresh outside air began to affect the clothing, the furniture, the textiles, immured in a dead chamber for almost four thousand years. Already he sensed with bleak uncertainty that what was done could never be undone. Softly, secretly, the molecules that had held their place through forty centuries were loosening their hold. "The life of the superlatively beautiful things around me," he recognized with a professional eye, "was now limited." Theft and cupidity in the decades to come would play their role. A grandeur was departing even as men strove to cling to the possessions of a god-king. In the end Howard Carter and Lord Carnarvon, the discoverers, would be at odds and die unhappily. Not even the protection of a museum, Breasted, the greater Egyptologist, knew, would long suffice to defend the property of a Pharaoh.

Men should discover their past. I admit to this. It has been my profession. Only so can we learn our limitations and come in time to suffer life with compassion. Nevertheless, I now believe that there are occasions when the earth tells our story just as well, when the tomb should remain hidden, the dead man masked in jade be allowed to lie sleeping at the temple's heart, and the temple, in its turn, repose beneath the crawling buttresses that mark the encroaching rain forest.

In the year 1975 twenty-one people died in an air crash at the great Mayan religious center of Tikal in Guatemala. Strange, is it not, that twenty-one tourists born over a thousand years after the fall of the old Mayan Empire, and only aware of it because of the archaeological excavations of the last two decades, should be drawn to that spot and die—die on a Guatemalan airfield specifically erected to draw the curious to the ruins. "The past is as open to development as the future," John Meerloo once observed. Here was a perfect example. Twenty-one people died who might otherwise have gone their separate ways in life. They died from curiosity about an alien city supposedly dead many centuries ago, but now resurrected from oblivion. Ten years earlier the past could not have extended such an arm into the future. Dead gods could not have fed once more on living victims.

I pretend no moral judgment. I only say that to tamper with the past, even one's own, is to bring at times that slipping, sliding, tenuous horror which revolves around all that is done, unalterable, and yet which abides unseen in the living mind, so that it may draw disaster from the air, or make us lonely beyond belief. I have had such an experience and it began, I am sure, in Egypt, though I have never dug there and indeed was an impressionable graduate student when I first encountered the story in the archives of the University of Pennsylvania library.

It was a long time ago that Herbert Winlock of the Metropolitan Museum of Art made the find at the tomb of Meket-Rē' at Thebes. No one, except perhaps James Breasted himself, would have had the sensitivity to see it as it was. The episode took place around 1920, just after the first World War, and some would have refused the concession to dig, because the noble's cenotaph had long been rifled. Even Winlock has admitted his despondency over the unpromising empty corridors, the piles of worked-over debris, the tunnels abandoned to bats. Men robbed in ancient Egypt just as they rob today.

But men of power in the Nile kingdom believed with a touching simplicity that you could take everything with you in miniature: your little fish pond of hammered copper, that would hold water, your spotted cattle, your pleasure boats, musicians, serving girls, even yourself in effigy. It was a more sophisticated version of what I had seen by the side of the dead child.

We, men and women of all ages, have never really accepted the fact of death, any more than after forty years I can accept the fact that the frantically running mongrel beside a vanished freight has really gone. Great kings, as in that dreadful grave at

(97)

Ur, have been known to take their living servants and concu-
bines with them in a single holocaust to which the doomed
came gladly to accompany their master. In Egypt, however it
may have been in the beginning, it was the miniature that assumed the semblance of life while the living were spared.

It was a gentle game, a game of childlike make-believe. There
were wood carvers whose duty it was to create the pleasant
palace and the precious palms, everything to insure that life
would go on with no loss of status, that there would be an oars-
man at the pleasure boat throughout eternity.

I thought, turning the pages of Winlock's report over a decade later, how enticing it would be if they truly believed, there
in civilization's dawn. If the funeral toys *really* comforted, if we
could still, in some magical manner, be transported to where
little figures in wood forever defeated the sting of death. I wondered, grubby student among old books, if they genuinely
meant it, if they had known a secret I had tried in a cruder, un-
skilled way to reproduce with gilded crosses. How similar we
all were across the millennia. Herbert Winlock convinced me
by the power of his own imagination that something *did* linger
in the little carvings, that men *had* believed in the miniatures
they hoarded against the moment of their deaths. Indeed in
the instant of discovery he had seen them alive.

His find came almost miraculously in an utterly gutted
corridor before the entrance to the vandalized burial chamber.
Meket-Rē' and his treasures were long gone. But Meket-Rē',
like the child with the gilt crosses, had had a secret. In a crack
of the rock floor Winlock's foreman had noticed a trickle of
sand ebbing away. Men had striven with matches to see through
the dark fissure. Finally Winlock had been summoned. He
brought a flashlight.

And there, he wrote, as his torch penetrated the darkness of
four thousand years, "I was gazing down into the midst of a
myriad of brilliantly painted little men going this way and that.
A tall slender girl gazed across at me, perfectly composed, a

gang of little men with sticks in their upraised hands drove spotted oxen, rowers tugged at their oars on a fleet of boats. . . . And all this busy coming and going was in uncanny silence, as though the distance back over the forty centuries I looked across was too great for an echo to reach my ears."

Meket-Rē', unlike myself, had successfully evaded the man with the scythe. His sarcophagus was gone, his real life treasures were gone, his halls lay in ruins, but with a child's mind like my own, though infinitely more astute and powerful, he had outwitted the vandals. In a miniaturized chamber cunningly concealed beneath the floor, his very self, his estate, his ladies, his fish ponds, and his cattle flourished while Rome and the Dark Ages rose and fell.

He had possessed a remarkable sophistication and the wise mind of a child. "Curse them all," I could hear his voice across the centuries. "I know the sly officials and the tipped-off robbers will come, but we, we, myself and all I loved, will live on beatified, invisible, while storm and violence trample on the floor above." What was the secret incantation for the transference, I wondered desperately. What had Herbert Winlock thought in that instant when all seemed living? Or had the little figures really existed as his first excited glance implied? Were they stunned into silence by the shattering light of modernity? No doubt the discovery was a great one, but I suspect, if the choice had been allowed me, I would have reacted with the guile of Meket-Rē'. I would have betrayed my profession, sealed off the cavity, and prayed that a second incantation had been held in reserve against accident in that tiny chamber. Not many years would pass, it so happened, before the modern world would be transfigured under my living eyes as was the past before the torch in Winlock's hand.

Perhaps there is no precise language in which to express such thoughts. A man comes into life with certain attitudes and is inculcated with others of his time. Then some fine day, the kaleidoscope through which we peer at life shifts suddenly and everything is reordered. I am sure something like this happened to Winlock as he stared into that tiny cavern, but the spoil, immobilized the moment the light beam struck it, was gathered up by dedicated men. Herbert Winlock, like the able professional he had proved himself to be, went on to become Director of the Metropolitan Museum of Art. Rumor has it that in the end he was unhappy, perhaps would have preferred to linger on in Egypt. He is gone now and that was very long ago. I will not vouch for his afterthoughts.

One does what one's time dictates, does one not? One does what one is ordered or expected to do, not necessarily because one is a coward but because one is a trained, conditioned professional. A soldier, really. The unexpected event, if it ever comes, leaves one unprepared and fumbling.

One exists in a universe convincingly real, where the lines are sharply drawn in black and white. It is only later, if at all, that one realizes the lines were never there in the first place. But they are necessary in every human culture, like a drill sergeant's commands, something not to be questioned.

Winlock happened to be an archaeologist who saw the lines shift and dissolve. A blink at the right moment may do it, an eye applied to a crevice, or the world seen through a tear. Then, to most of us, the lines reassert themselves, reality steadies out. It is better so. Every now and then, however, there comes an experience so troubling that the kaleidoscope never quite shifts back to where it was. One must then simply deny the episode or adjust one's vision. Most follow the first prescription; the others never talk.

I have mentioned that there was a kind of frenzy that seized men about the events of the terminal ice as they were interpreted in the late 1920s and early 1930s. There were no precision methods of dating then, and many bitter feuds arose among scholars. A kind of wild surmise, later to be proved correct, proclaimed that man had a much greater antiquity in the Americas than the conservative "establishment" would at that time accept. It is all in the books now, but in the days of the dust bowl beautifully flaked chalcedony dart points were lying ever more exposed in sand blowouts as the topsoil blew away. There were rumors of such points found within the fragmented skeletons of extinct, long-horned bison, and great contention existed as to when the creatures had vanished.

Dead lakes, like that I had earlier prowled over in the Mohave, were explored in New Mexico, sometimes with spectacular results. It was natural that the desert west of Carlsbad cavern should come under attention, just as did that dust-filled shelter that had preserved the culture of a later time on the slope beneath the Texas mountain.

The terrain around Carlsbad in those years had certain peculiarities. Carlsbad cavern itself is one of the largest in the world but in that time, though open to tourists, its full extent was unknown. Seeping underground water over long ages can produce enormous mile-long passages in limestone deposits. In the course of ages these streams can descend from one level to another, leaving the upper caverns dry. Holes may twine off in all directions. Huge chambers may form and long ages later collapse. Sudden drops into lower abysses may endanger the unwary. The Carlsbad country, far beyond the known scope of the cave, was as riddled as a termite nest. Sometimes on the flat desert floor one might come upon a fissure whose depth was anyone's guess.

Now the archaeologist is primarily interested in shelter caves where daylight is accessible. Shelter from the elements, in other words, but not the awesome underworld into which early man might wander and be lost. Nevertheless, sinkholes, which are really collapsed sections of cavern roof, have sometimes trapped and held extinct animals out of a time dating to man's first penetration of the region. So the hunt went on, decade after decade, and, of course, amateur explorers have never been able to leave dark openings alone. The disastrous fate of Floyd Collins was a well-publicized example.

THE CREVICE AND THE EYE

Inevitably word was brought to us by one of these amateurs that there was a cave in the back desert where animal bones, possibly those of extinct camel, had been found in the slide of an overhead sinkhole collapse. Every report of this nature had to be checked, though most were bound to prove baseless. We went by a way I would never be able to retrace. A good-natured Catholic priest guided us to the site and we were provided with what we were assured was a reasonable sketch of the passages explored by the original visitors. The map, we were to discover only later, was correct up to a point, but it failed to provide important information about other side passages.

The site was a curious one, though we came well equipped for the first part of the adventure. A rather inconspicuous tube, a kind of giant wormhole, descended into the depths of a low hill. Here we paused to examine our map while the jolly fat priest explained things to us. We would go down the tube carrying a rope ladder intended for such purposes. At the end of the tube was a sheer drop into nothing except for a giant stalagmite to which the rope ladder might be attached. Many feet beneath us was a huge chamber into which we were really to enter through the ceiling. From that chamber, our hastily constructed map informed us, separate corridors radiated. We were to take the one indicated and follow its course to the rock slide and the suspected bones.

In that spot upon the ceiling, after a tortuous, winding descent, the priest paused. All good priests, I thought briefly, have a right to pause above unfathomable abysses. "I am too heavy for your ladder," he explained. "Rope it around this pillar and I will stay up here and wait." We could not complain. He was giving us the flashlight. He would have to sit in the dark alone.

I shot the light over the edge. The chamber was enormous, the climb, even with the ladder, not inviting. We braced things as we could and I went over the edge, followed by my companion. Four separate corridors extended in opposite directions from the main chamber. We tried to follow the trail delineated by the map. We went deeper and deeper until we were crawling over tiny stalagmite crystals that pricked our knees like needles. We could no longer walk. We were creeping now, and the air seemed close and very bad. Finally, by mutual consent, we halted. The batteries in the lamp were beginning to show signs of stress. "Shut it off," I said. We sat for a moment in stifling silence.

I knew just one thing in our favor, though not really pleasant. Up where the priest waited there was only a single way out. He could not get lost even without a light. But without a light how long would he sit there? How long would it take him to make his way up that tunnel in the dark and drive to town for help? Our tension mounted. Men could panic with claustrophobia under these conditions. They sometimes crept on, losing themselves in a labyrinth when they should have waited for help where they were.

The rested lamp was relit and cast behind us. All the shadows were different and remained inscrutable. Our lungs labored in the low, flat dark. We could not recognize the way we had come. Stay, I thought at first, stay and engage in as little exertion as possible. But how long will it be before the priest knows we have vanished? He knew we had an indefinite distance to go. I respected his courage sitting alone up there in the dark. I cast a ray of light behind us. "Look," whispered my companion in timid hope, "we are the first people ever to come this far. We have been climbing over those tiny stalagmites, breaking them. Maybe we can follow our own path out of this place."

At this point, in the still, close air, we made our decision. We would not wait. Slowly, husbanding the light, we began to work our way back through the fragmented crystals. Finally we came to a place where a long forgotten mud torrent from the desert floor had cascaded down a corridor. There were footprints in it—undoubtedly those of the map makers. Slowly we traced them backward with much trial and mistaken effort. After what seemed an infinite period of time, we emerged once more into the chamber through whose roof we had descended. Warily we flashed the light over the still-hanging ladder, flashed it again toward the far-off ceiling. A voice floated down, patient but concerned. "Say, you fellows," called our friend, the faithful priest, "time goes awfully slow up here without a light. I was beginning to wonder."

We looked at each other, each man, I suppose, with his own private thoughts. "All right," I called, "give us a minute, we're coming up, but the light is weak. You seem awfully small. Just steady the ladder. Everything is under control." It ought to be, I thought, having a priest at the other end of the rope.

Slowly my companion ascended, swaying and twisting in the tiny light while I steadied the ladder from below. Then I followed. After catching our breath we stumbled on our knees up the tunnel, dragging the ladder with us. There was nothing much left of the light, I thought, groping. It was going out. We must not make that mistake again.

But the mistake was already made. In the end, you see, my angle of vision was twisting in a way opposed to Winlock's; I was staring from the other side of the crevice, from the kingdom of Meket-Rē'. It was we who were reduced to pygmies. By the time I stood at the cave entrance I was looking at life, at my companions, at the traffic below on the road, as though I had just arisen, a frozen man, from a torrent of melting ice. I wiped a muddy hand across my brow. The hand was ten thousand years away. So were my eyes, so would they always be, and still, like Winlock, I did not find a way to speak.

The modern world was small, I thought, tiny, constricted beyond belief. A little lost century, a toy, I saw suddenly, looking upon our truck and pretentious archaeological gear with that stunned insight which had overtaken Winlock in the tomb of Meket-Rē'. Winlock had looked into a cavity and seen the past in motion, and stilled it with his torch. I, and here I looked curiously and distantly upon my associates, had arisen from some kind of indefinable death amidst stalagmites and glacial mud ten thousand years removed. That mud was on the battered flashlamp as we neared the portal. It almost dimmed us to oblivion.

"We are dwarfed," I muttered to myself alone, "the tiny projection of a lantern show."

I have never again seen men so minutely clear, though I climbed with the others into the pickup truck and held my peace. "Dwarfed," I said again under my breath, seeing ourselves moving with tiny gesticulations across an infinite ice field into whose glare we finally vanished. It was like a glimpse through the slitted bone with which Eskimos protect their eyes from snow blindness. I have never had occasion in the years since to think upon us differently. Not once.

CHAPTER 11

The Growing Shadow

A FEW days ago, meditating upon the past as older men will, I came upon a jotting in a notebook. The date was November 17th, 1957. It read, somewhat cryptically: "In the midst of the faculty party: the rain, the rain, and loneliness of my student days brought back by a casual word." What was that spoken, terrible, evocative word? Memory no longer informs me. Perhaps its power lay in the gaiety, the circumstances. My first winter in Philadelphia was not happy. I was too poor to afford a room in the graduate dormitories. I lived alone in a room provided by a family some little distance from the university.

It was true that I had furthered my acquaintance with Ricky and, in turn, with one of his Japanese friends, whose nickname was "Duke" or, sometimes, "Tachi." Still, they lived in the campus dorms. When evening fell I looked out upon rain falling endlessly under the street lamps at a nearby corner. I was homesick. The high plains are cold in winter but they were sunny and unpolluted in that time. There was little in the way of diversion in Philadelphia because funds had to be husbanded carefully. I was just barely getting by; furthermore, I was not adjusted to the climate. Colds assailed me. The rain seemed never to cease in that autumn of 1933.

A short time before I was to give a seminar report which the entire professorial staff would attend, I came down with a severe cold which spread into my right ear. If I had had money or even a sense for city survival, I would have sought medical aid. As it was, I found myself trapped, over a long week-end, with a throbbing ear, in which the drum finally broke. It was a wonder that my mastoid did not become infected.

I gave my lecture, but to my horror I was failing to hear some questions directed at me from the back of the room. I was not heartened to hear later that Speck had remarked in his gruffest manner, "He's deaf as a post." Coming from him this could only mean impending doom. I counseled with my friends. Duke was a biology major, aiming toward the Medical School. My fearful memories were those of the way I had been treated in that shabby little dispensary in Lincoln. I was too ignorant to know that Penn possessed one of the finest medical schools in the country. "Look," my friends advised, "when it first came on you should have gone to the outclinic of the University Hospital itself. Where did you think you were? In Pine Ridge? You're a properly registered student entitled to medical attention. Come along, Duke will show you."

I was taken to an efficient-looking waiting room where I was presently received by a skilled and able young intern. "Well, the drum is ruptured," he diagnosed. "The pressure has been relieved and that is perhaps just as well. We've got to get a mass of wax and hardened discharge out of there. The hearing should then pick up by degrees." He gave me some oil to soften the interior mass and asked me to return later. I was decently, expertly treated. Things came about as he said. I never inquired the precise degree of my hearing recovery. The point was, I could hear. Speck and the other professors quit eyeing me dubiously. Only long, long afterward, fifteen years to be exact, did I have some cause to wonder about that ear and the effect of winter rains in the great coastal city.

None of us in that time was enjoying the diet of today. Duke,

Ricky, and I generally met to have breakfast at a cafeteria run by a wily old huckster in whose food we occasionally detected pieces of rope, tin, or other indigestible objects. When all of this became too intolerable we would march firmly by the cafeteria in full view of its owner. If we persisted in this, the shrewd old operator, well aware that students talk, would come outside and invite us in for a meal "on the house." I often wondered if some of his spicier preparations came out of the neighborhood garbage cans, particularly after I had experienced the rope episode, but the old man had his softer side and was drawn almost paternally to Duke, who seemed to know how to manage him.

There was a Chinese restaurant farther away on Market Street that we visited in more affluent periods as the mood might strike us. It was the old Nanking, now long since gone, the victim of the changing nature of the city. Men of all races foregathered there and the food was excellent. Generally when we went, it was as a group of several, in order that we might share dishes more economically. There I first learned to wield chopsticks, determined as I was not to set myself off from my Oriental friends. To Japanese-born Ricky, of course, all this came as naturally as his banter in Japanese.

Slowly the sullen sunless winter dragged by. Speck, I discovered, had a way of vanishing with the northbound geese. His graduate classes never had a precise date of termination. Instead, he grew increasingly restless as spring advanced. Finally the office would be empty. There was a family home at Gloucester if he was not going farther. As for the students, in small comfortable numbers at that time, they, too, drifted away by degrees, leaving papers and setting forth upon field projects greatly valued by the department. Funds were dreadfully sparse in those years. Mostly, one had to seek one's own arrangements. I made mine, but that is another story.

In my second year Duke, Ricky, and myself acquired a large room together in the International House at 40th and Spruce, an old mansion then owned by the University. The house has long since been demolished, as has the hospital in Lincoln where I was born, and the grade school I attended. The wreckers, it sometimes seems to me, always follow fast upon my heels, but then what else should an archaeologist expect? It is just that as one grows old one wishes to go and look where one should not look. Only the sun-warmed stones, if any remain, will set things right—the stones that radiate warmth, but never speak, though there may be names upon them.

It might be asked why we were able to acquire and share this huge old room in a building devoted to foreign students. The answer is simple. In those years of the depression the number of foreign students had declined disastrously. The house had to be sustained, and Ricky, myself, and Duke, an American-born Japanese, were amicably interracial in our friends and connections. By chance we acquired the master bedroom of the old mansion. There was a giant wall safe implanted by the fireplace. The room proved a far more happy solution for us than we had anticipated. There was a good deal of space available for books, and, in addition, any spare moments could be spent twirling the dial of the huge old safe, in an attempt to solve the combination. It stubbornly resisted all our efforts but the unsolvability of the problem was, in itself, a sheer delight on damp days. We interspersed our efforts with dreams of abandoned wealth or historical documents of great value. The combination of that safe was lost in another century.

The year was one of the happiest of my life. We never quarreled and the room was somehow warming to visitors—Chinese, Indians, Filipinos. Our visits to the Nanking restaurant increased, as did our numbers. It was a strange time in which to be alive because of the many paradoxes it presented. As we all dined amicably together—black, white, yellow—terrible forces were at work beneath the surface of things. Jobs were few, yet I could walk after midnight from the homes of friends and pass safely through neighborhoods from which I would not now expect to emerge alive.

Some of our Oriental acquaintances came from rich families that would be forced to flee or perish before the Japanese invasion, or, if not that, would die in the post-war convulsions from which emerged Red China. The son of the mayor of Shanghai was among us, as was the son of a notorious Philippine gambler. I wish I could name them all, because in my later experience of embittered or truculent minorities, I was never again to encounter the genuine peace I experienced among these individuals whose national governments were already making the first moves on that chessboard whose pawns were to be swept away in millions.

I said I wished I could name them all. As my departure approached, I was presented at the old Nanking with a Modern Library edition of Thomas Mann's *The Magic Mountain*. Inscribed within it were some fifteen foreign names. I carefully put it away in my aunt's home before she abruptly sold the house. It was never returned to me. Sometimes as I scan with aging eyes today's newspaper headlines I wonder if some of the names I see are those of my old companions, or if they lie in nameless graves in jungles, or in Micronesian waters.

"Never make the mistake of underestimating the Japanese Navy," Ricky, in a serious moment of discussion, had once said to me. He had been right, more right than our own military intelligence, but, of course, no one had listened. And where is Ricky? He despaired and turned to high-school teaching. I last heard from him in 1938. The war was not too far away. Saburo Kitamura, the famous Japanese doctor who was honored in Tokyo in 1975, was among the visitors at the House, but not one of our intimates. He was already far advanced on his medical career.

I have said, and I reiterate, that it was a time of peace among us—at the House, at the interracial eating places we frequented.

Chou Li Han, one of the Chinese graduate students in the department of anthropology, decided to take us to his own favorite restaurant in Chinatown. Chou Li Han was a reserved, serious scholar who stuttered a little—at least in English. We had no idea from whence he came, what his origins might be, or where he might be going.

As we entered the door of this elegant establishment, Chou Li Han clapped his hands sharply. The proprietor bowed, waiters scurried. Obviously Chou Li Han and his guests were something special. It was there I had my first experience of bird's nest soup. Like most Americans I expected to see sticks in the soup but was quickly set right by Ricky. Ricky, who was on very good terms with Chou Li, nudged me and said, "You'll never taste anything better in your whole life." His judgment was correct.

There was just one thing disheartening to me about these Oriental banquets. I was not a linguist and I had to wait upon my Chinese and Japanese friends to order the rarest fungi, the strangest culinary dishes, in two old and highly ceremonial cultures. By now I used chopsticks so much like a native that I was once asked by a highborn American diplomat in what part of the East I had grown up as a child. "Nebraska," I was tempted to venture, but I answered gravely with the truth of the matter.

Violence was beginning to stalk the world. The first harbingers of the storm in Nazi Germany were beginning to arrive, not only Jews but blond Aryan anthropologists who could not stomach the racial doctrines they were expected to disseminate. They arrived almost penniless and, such was Speck's fame, sought him out immediately. As among all refugees, there were the good and the bad. Some were grateful for what little aid faculty and impoverished students could muster; some drifted on to Latin America. Some found posts but could not adjust to American ways and bit the hand that fed them. As student onlookers we sensed all this but did not realize the fates that were to befall us. Chou Li's dinner was, as it turned out, the end of innocence.

The whole matter turned upon our visit to the provincial capital of what was then the greatest power in the world, the United States of America. Corpus, a Filipino, owned the car, and the rest of us came along to see this place where government and justice were dispensed. There was one of the Chinese, an Indian or two, Duke and myself.

We saw the outside of the White House, peered at assorted monuments, and visited an art gallery. During the afternoon we had some minor car trouble which took time to overcome. It was late enough to make us consider the advisability of staying overnight. We went to several hotels which, surprisingly in that depressed era, always seemed to be full the moment the desk clerk caught sight of the faces behind me.

"The YMCA," someone, perhaps Duke, finally muttered. As spokesman for the group I was beginning to find difficulty in looking my friends in the eye. We found the YMCA, the Christian organization which in my time was supposed to keep young men the right path. Naturally I had had no contact with it. Growing up the way I did, I had never been invited to join the Boy Scouts either. For the sake of my friends, however, I was willing to try. "Sorry," said the clerk.

"I know, I know," I completed for him. "The town is unusually full. There is a dearth of rooms even in the Christian Association."

"You know," he said, "you're absolutely right."

In the car we held a conference. This was before the day of the professional motel. Private citizens with a few rooms to spare sometimes put out signs for motorists. "We'll go out into the suburbs along our route home," I argued. "We'll find something."

We did. A respectable-looking large house carried a sign, "Tourists welcome overnight." Corpus slumped miserably behind the wheel. I went up and knocked at the glass door, Duke a little behind me. A pinched-faced woman with the relentless eyes of a turkey vulture came to the door. I explained our need for rooms. Her eye traveled behind me, focused upon Duke. "Is this your man?" she asked.

Never in America had I previously heard that expression. I thought it was something only to be read in English novels. I was indeed still very naive.

"Ma'am," I said, "he is not my man. He is my friend." The door smashed shut as though glass would be spewed over us. I went down to the curb.

"Fellows," I said slowly, "I guess we are in for a long drive." We speeded; we felt like speeding. The Nanking was still lighted when we arrived.

After we were seated and had selected from the menu, I looked up. I looked at each face in our little party. I no longer felt white any more than they did. I was filled with the rejected fury of two continents. "If one man," I said, "can apologize for a nation, his nation, I apologize." They were all softly gentle with me. But long, long after, long after World War II was a memory and a new third world grew menacing and mocking in the United Nations, I thought upon that great world capital and how many tiny acts of folly by pontificating clerks and old women steeped in malice had helped finally to usher in the nightmare world of the present. History is what it is; men's violence falls alike upon the guilty and the innocent. Perhaps our little precious comradeship was doomed from the beginning. I do not like to think so.

Duke, citizen of the United States of America, went home in 1936 to see his family and try to arrange the transfer of an inheritance. Hardly had he arrived in Japan before the Japanese government made it impossible to transfer funds abroad. He lingered helplessly with his relatives while the storm clouds gathered.

In the end the clouds rolled over Honolulu. Somehow, with the help of his relatives, Duke survived. In the fire bombing of Tokyo, police records were destroyed. Duke fled with family help into the countryside. His story is too lengthy to narrate here and I possess only the bare outlines. Let it suffice that an American intelligence unit located him. As a biologist he aided our medical investigating teams at Hiroshima. He returned to the States and sought me out. Recently, two graying men, we sat at lunch together. "Tachi," I said, "you're retiring, what now?"

"Eiseley," he ventured carefully, "my sister and her family have an unused country house in a national forest. It is very beautiful and quiet there. I am going back for a visit. They want me to stay now . . . with them. Here . . ." he paused. I knew what he was thinking, the man lost between two cultures, the man whose vital years had been consumed by war and its ghastly memories, the man who had not married, the man who had been distrusted in Japan, but who, if he had stayed in the United States, might have been equally under surveillance. I thought of that night long ago in Washington. "Duke," I said, "let me hear. I wish we could go together."

"That would be nice," he said. "That would be less lonely. You see, I . . ." The serene fatalism of his race immobilized his features. He slipped softly out of the door. I had a vision of him sitting alone like an ancient sage beside some mountain pool. He had once been different, laughing, eager. Perhaps he lived now with the dead of Hiroshima. I had no right to ask; his calm was beyond my probing.

"All those others," I had said to him at one point in our luncheon, "those who signed in *The Magic Mountain,* I never heard of again except for one." "And who was that?" he asked. "Chou Li Han," I answered. "Remember how eager he was about his research? Someone came back from China before the Reds took over. There was a story that Chou Li had been eaten by a tiger in the Tibetan highlands."

"We go a long way to our deaths," Tachi had meditated. "I suppose that alone has immortalized him in your department."

"No," I said, "the truth is I am the only American who knows about him now. He was such a quiet little person to be eaten by a tiger. It seems odd, somehow fated." Tachi had nodded but his eyes had been far away. He had stood where a city had perished.

That night I tossed and thought of my own departure by plane in that time so far behind us. There were none of the great moving ramps then, no complex jockeying for take-off. Just a plane and a winter flurry of snow. I had peered compulsively out of the window by my seat. No one was there, but I kept staring as though expecting someone. People were still wary of flying. Sometimes if snow threatened one had the cabin to oneself.

Actually I was flying in an attempt to reach my grandmother's bedside before her death. The motors roared till the cabin vibrated. Then the plane began to speed down the runway. It hurtled toward the point of no return, reached it, leaped upward. I lay back, still troubled, as though I expected a companion.

I thought of Han toiling relentlessly on and on through the Himalayan solitudes. A great yellow-eyed beast had come slowly to meet him in the snow. Snow. Snow swirling before a prison gate. Snow as I had known it howling across a high-plains cemetery in winter. Snow sifting over all those years until even faces once known were now fixed, impenetrable. I strove to remember them. They were sealed, each one, in a crystalline silence.

"Would you like a drink?" the stewardess had asked, pausing by my seat. Somewhere far off I had sensed a little yellow man looking upward as though he saw me—Chou Li, the last signer of *The Magic Mountain*. "I would," I had answered. The propellers were swinging into the rhythm that would cross a continent. They never faltered.

"There is a blizzard ahead," the girl had confided. There were only a few on board and things were more informal then. People needed to sustain reality.

"I was just thinking," I had said as I took the drink, "thinking about the snow. One always does, doesn't one, on night flights this time of year. Snow shuts things out, or it lets one see . . ." I sipped again. The stewardess had looked at me curiously. "No matter," I had said and drained the glass. "Just a dream. A man in the snow. Someone I knew. A dream."

Now after the years, Tachi had returned with that voice from behind the wars. Much lay hidden between us. We were both grey. My book was lost, with the signatures. My friend was leaving, not to return. Slowly I looked at the lacquered box with the red dragons that had been his pre-war gift. I drew it closer, fumbled uncertainly, and opened it. There was a photograph of three scarcely recognizable youths, my roommates and myself, standing on the steps of a vanished house. Goodbye, Tachi, I whispered, but it was a farewell for all of us—the signers in *The Magic Mountain*. Our world was gone.

CHAPTER 12

When the Trouble Comes

I N the fall of 1936 I belatedly entered a crowded coach in New York. The train was an early-morning express to Philadelphia and what I had been doing in New York the previous day I no longer remember. The crowded car I do remember because there was only one seat left, and it was clearly evident why everyone who had boarded before me had chosen to sit elsewhere.

The vacant seat was beside a huge and powerful man who seemed slumped in a drunken stupor. I was tired, I had once lived amongst rough company, and I had no intention of standing timidly in the aisle. The man did not look quarrelsome, just asleep. I sat down and minded my own business.

Eventually the conductor made his way down the length of the coach to our seats. I proceeded to yield up my ticket. Just as I was expecting the giant on my right to be nudged awake, he straightened up, whipped out his ticket and took on a sharp alertness, so sharp, in fact, that I immediately developed the uncanny feeling that he had been holding that particular seat with a show of false drunkenness until the right party had taken it.

When the conductor was gone the big man turned to me with a glimmer of amusement in his eyes. "Stranger," he appealed before I could return to my book, "tell me a story." In all the years since, I have never once been addressed by that westernism "stranger" on a New York train. And never again upon the Pennsylvania Railroad has anyone asked me, like a pleading child, for a story. The man's eyes were a deep fathomless blue with the serenity that only enormous physical power can give.

People on trains out of New York tend to hide in their own thoughts. With this man it was impossible. I smiled back at him. "You look to me," I said, running an eye over his powerful frame, "as if you were the one to be telling *me* a story. I'm just an ordinary guy, but you, you look as if you had been places. Where did you get that double thumb?"

With the eye of a physical anthropologist I had been drawn to some other characters than just his amazing body. He held up a great fist, looking upon it contemplatively as though for the first time.

"You noticed that?" He laughed without any show of embarrassment. "Then how about this?" He swung his hairy right hand up and around. The backs of the nails were raised and thickened like the claws of an animal. Furthermore, they ended as sharply as claws. Tiger claws, I thought with a little shudder, except that they were not retractile. "In a fight . . . ," I considered reflectively.

"The sea is a rough place," said my huge friend as though following my thought. "And so is the shore. Us Rileys, that's my name," he rumbled, extending the hand with the double thumb, "are all like this. We just bust out all over." For a moment I saw practically that a double thumb in a rough and tumble could be used to jamb into two eyes at once.

"My name is Loren Eiseley," I said, taking the great paw. I did not try to explain about being a graduate student. Somehow it did not seem appropriate. "I know it's a hard name to remember. Don't think you have to try."

"Of course I'll remember," he said in a voice that had been

used to outshout hurricanes. "My name is Tim, Tim Riley," he repeated. People down the aisle turned and stared at us. Once more that fey sense touched me that the seat had been held for me alone. "Okay, your name is Loren. Now listen. My old man —and I haven't seen him in years—is a police captain in San Francisco. If you're ever in trouble there, bad trouble"—I could see the tiger claws grip unconsciously on the back of the next seat as though to strip the upholstery—"tell him Tim Riley is your friend. Tell him"—and again the huge claw clutched and relaxed rhythmically—"tell him I'm still alive. Tell him—" Here he paused and settled back in his seat, holding up his powerful hands in a suddenly puzzled fashion. "Jesus, time passes different out there." He waved vaguely in the direction of the North Atlantic. "I just had a thought. Maybe the old man's retired or dead.

"But ask anyhow," he reiterated with the philosophy of the world I had once known. "It can't make it worse for you when the trouble comes, and cops are always trouble. Ask for John Riley. You can't mistake him. He's big, like me. You'll know him. He'll fix things if they can be fixed. Only tell him, see, it's for me, Tim, and I'm still alive after all these years. Tell him it would be a favor."

"Sure, Tim," I said. "I know how it is. There's always the trouble sooner or later. But now you tell *me* a story. I'm a landsman. You wouldn't be interested in that."

The claws like those of a tiger cub began once more to work rhythmically over the seat in front of him. "The docks, you know," he began. "The docks in Frisco."

"Of course I know," I said honestly, out of an earlier time. But there were no jobs then unless you had connections. I had merely looked longingly at the ships and ridden on the ferries. I judge him to be ten years older than I, but I could have been wrong. It might have been his size or the sea leaving its marks.

In retrospect, I am aware of a question of veracity in my treatment of one of the most physically unusual men I have ever met. Most of us go through life thinking merely of people with variable features, but all essentially alike. The physical anthropologist, perhaps alone, is the most conscious of human differences, strange mutations in the normal run of things, inexplicable emergences, atavisms, all that difficult entangled thread that produces successive generations. The hidden alphabet of life draws some characteristics into reality and suppresses others. As the French biologist Jean Rostand has aptly put it, "No man has a true counterpart."

Rostand was simply expressing in epigrammatic style the fact that below the existent men of every given generation there lurks an army of *potential* men. The world is never quite where we see it. In our germ plasm, perhaps mercifully beyond our knowledge, are hidden the freaks, the geniuses, the anomalies of tomorrow. I had an eye for these things, sharpened early by my experience with the life machine, the hatching trays I had been forced to paw over in my youth. I had learned that the machinery of life is gambling machinery bringing into existence both the beautiful and maimed. I had seen the reality in all its shapes, like dice throws on a green table. Moreover, I had come to know surprises, the one-in-a-million throw. I had learned never to underestimate the potential in favor of the actual.

For example, there is supposed to be, within variable limits, a certain symmetry about the human body, proportion, shall we say. Yet, sitting idly on a bus a few months ago, I observed across the aisle from me a creation I never expect to see again in this life. It was simply a woman in whom proportion did not exist. Her upper arms were too short in proportion to her forearms, her head too large for her body, her torso out of harmony with her legs. One thing, however, must be made plain. This was no product of thalidomide. This woman, if we stretch the human parameters a little, was normal. I doubt, except for perhaps some consciousness that she was not beautiful, that the woman in question ever thought of herself as a genetic anomaly. Yet most certainly she was. Her general appearance to the trained eye was that of a creature manufactured out of disparate, ill-assorted parts. The symmetry factor was lacking. Although one would have to label her, in some vague fashion, "normal," she existed in reality in a no-man's land of "parts." This she would never know.

Or take the spider-finger syndrome, arachnodactyly. Statistically it appears to have a greater incidence in certain inbred ethnic groups. The hand is tenuous, sometimes lengthy, but the fingers, while otherwise normal, remind one of something in the nature of an insect scampering over the page. Incredibly thin and elongated, they are spider fingers in very truth. Lay people occasionally observe these things half-consciously but they are frequently unaware of their rarity, the storehouse of potential lying below the actual from which these oddities are drawn. They do not know that in their own germ plasm may linger things as unique. Indeed one may occasionally look back upon a fragment of the past, as in the shape of a huge brow ridge, or even see the unknown features of our far distant progeny prematurely peeping into existence.

"Tell *me* a story," I again chided my friend with the tiger nails as the train picked up speed out of Newark. "The sea is your trade, isn't it?" We had passed the waving grass in the Jersey marshes. I watched the clawed hand pluck the cushion and knew a time ages ago when men had had nails somewhat equivalent to these before our claws had been spread delicately flat as part of a different sensory system.

"When I was a kid I hung around the docks," he said. "My old man was a cop, like I told you. He was big and a fighter. Those were tough days."

"He beat men into line," I said suddenly, wryly remembering the words of the gunman with whom I had ridden over the Hump.

"Yeah, yeah," repeated my companion. "It wasn't easy to walk that beat. And the men he met weren't easy, but, man, I'm telling you he was a fighter. In the end he ruled that precinct like a king. Nobody like him. I wonder . . . ?" He paused.

"No," he responded to my unasked question. "I never wrote. I don't know why. I ran away when I was fifteen. The old man was all right. No trouble there. It was just that we were all big. Nobody to stand up to us, I guess. It was a way of life." The blue eyes, impenetrable as the sea, studied me a moment. "We went our own ways soon as we could. Mother was dead, Pa rulin' the docks. He was all right or I wouldn't be giving you his name as an ace for trouble. But you see, he didn't like weakness, and I didn't like it. I near killed a kid made fun of this hand. I used it on him."

The rhythmical shredding began once more on the seat before him. "After that, so I wouldn't cause the old man talk in his precinct, I shipped out as a deckhand. I learned the hard way, bottom to top. Now I'm a marine engineer. But I never wrote back. I've lost track of them all. It was just our way when the time came to leave the nest. The old man would have understood." He repeated again, as though it were a ritual, a way of keeping in touch, "When you're in trouble in that town, ask for Captain Riley and tell him I sent you. He swings weight and it could help you."

I wondered briefly how many times in how many forecastles out at sea or in waterfront bars he had said, "When the trouble comes—and it will—ask for my old man. Tell him, I sent—"

We both held our thoughts for a moment. The train was picking up again beyond New Brunswick. I wondered briefly if any man had ever asked in the bullpen in Frisco for Captain Riley and had had the nerve to say, "Your son Tim told me to ask for you." I rather doubted the message had come through.

"What are you doing in Philly, Tim?" I asked, breaking for once the etiquette of the wanderer.

"It's like this," he said. "There's a strike and a picket line. The company wants me to get a freighter out to sea. I'll get through the picket line all right, but we're mighty short of hands." He shook his head.

"I reckon your company picked the right man," I said with genuine deference. "Mind if I see the hand?" I queried gently. "I happen to know something about these things." I thought I was running a certain risk but to my surprise he yielded it up like a trusting child. I felt the nails. They were rock hard, a sheer medical marvel. "I suppose," I said, "one could get something done about it now, but I reckon they've been useful." We grinned at each other like fellow conspirators.

"You're right, buddy," he said, and flexed the claws. "Useful in certain places and I'm big enough now nobody laughs twice." Again we smiled in sudden intimate harmony.

Then he changed the subject. "All the seas," he said abruptly, as though there were no longer time for a story. "Whorehouses up every alley, fights, drinks, the way the time goes by ashore, you know. Not like out there with the big engines needin' you. This shore stuff, the bought stuff, sure a man needs it, but, stranger, I'm telling you, it's miserable fun."

"Time maybe for something else?" I ventured.

"That's what makes it hard, Mister," he said. "I got a real girl now in New York, a nurse I met. I didn't want this run. We're talking about gettin' fixed up permanent, you know. Kids maybe." He sounded excited. "But then I get this order for Capetown. Christ, and those engines. They need care. They've been my life."

"Listen, Tim," I said, "write to her and come back. There's an end to this sometime."

"I haven't written anyone for years," he said, flexing the great paw in contemplation.

"Tim," I pleaded, "listen, just once. You've got brains. I'd be blind not to see it. You've come all the way up to the top. It's the sea that's got you hooked. There's a time to unload. Write the girl, write and come back." We were rolling into Philadelphia now.

(123)

"You didn't say what you do, you didn't tell me a story," he prodded.

"You wouldn't buy it, Tim," I protested. "I haven't got a job. I'm a grown man going to school at a university." I waited for his contempt. We were walking up the aisle now and then standing together on the escalator.

"Well," he ventured doubtfully as we finally reached the vast hall of the station. "Somehow, you didn't talk that way. Look, do you like what you're doing?" He tried once more. " 'Cause if you don't, you can come with me. I'll get you signed on. We're real shorthanded, like I told you. I'll show you the ropes myself. All we do first is run that picket line. I'll get you through."

A great surge of sympathy swept over me then, as well as hunger. It was as though we stood at the end of the world and someone was saying, "Come over, I'll help you into another universe." I was tired. Good God, I was tired, and there were no prospects of jobs even if one finished. All along the way I had seen my graduate fellows drop out—turn to high-school teaching, turn to anything but this meaningless treadmill of the depression. We stood there helplessly, the six-foot-five giant with the clawed hand who loved the sea, and a boy who was suddenly back listening to the stories read by a more literate mother of other children on a next-door screened porch high above my little bedroom in Lincoln. *Treasure Island,* and then those things discovered by myself, *Moby Dick, The Nigger of the Narcissus,* came flooding back to mind. I never thought of myself as inadequate to the occasion. The sea. A torrent opened in my head, the thunder on a thousand beaches. "Go now."

I steadied myself as in a giant wind. On the fourth floor of College Hall at Penn was that place of books in which one of the last great scholars of the turning century had his abode, the man who was basically as alone as myself, whose grandfather had perished with his vessel, lured by the false lights of wreckers on the Jersey shore. Frank Speck, who had guided and chided my graduate career. I thought also of my dead uncle who had set me on this road. Did I owe them nothing then? Sadly I looked up at the patiently waiting giant. "Tim," I said groping, "don't get me wrong. I'm not afraid of the pickets. I'd like like hell to go. I'd rather go. But I've got a debt to pay, not in money, but to people. Do you understand? I guess that's the story I never had time to tell."

We stood a moment more at the taxi rank to which I led him. He had engaged in too many partings over the world to make a thing of it. The tiger's paw touched my shoulder a moment. "Goodbye then, stranger," he said, already lapsing away from my own name. "I wish you could have come. And remember in Frisco when the trouble comes, ask—"

"For Captain Riley," I said, as though he would be waiting there for both of us throughout eternity.

"Right," said the big man turning on his heel. "Pier 10 on the Delaware, driver." He got in. I raised a lonely hand.

Just six years later men were dying amidst blinding steam in the engine rooms of torpedoed freighters off the Jersey coast, or leaping into blazing oil on the Murmansk run. I wonder if he ever wrote to the nurse in New York. Probably not; he was not a writing man. I wonder about police captain Riley and if I still must ask for him when the inevitable trouble comes. I suppose I must. The war is done now but so great was the public's concentration upon armies that few remember the vast tonnage sunk, or the deaths men died who serviced the immense engines below decks, sweating out the moment the torpedo struck.

In the end I stood up alone without friends or relatives and received my doctorate. I knew a good deal of ethnological lore from the Jesuit Relations of the sevententh century, about divination through the use of oracle bones. I knew also about the distribution of rabbit-skin blankets in pre-Columbian America and the four-day fire rites for the departing dead. And mammoths gone ten thousand years, I knew them, too. It was all happily irrelevant, but unlike today's students I enjoyed its irrelevance. My doctoral dissertation was entitled "Three Indices of Quaternary Time: A Critique." With this out of the way, I set out in the pursuit of a job.

Many years later a distinguished colleague and university official once stopped me on the walk. It was some time after the death of Frank Speck, my old mentor. "Tell me," he said, "was Frank Speck all that good? To me there always seemed a touch of the charlatan about him."

I paused a moment in serious thought. "He was basically an Indian," I said, turning the answer over in my mind. "He thought like one. He was reared by one. He believed in things, ways of life, that neither you nor I believe in, and perhaps this is not necessarily to our credit. No, he was not a charlatan, if you mean to derogate his importance. He was a genuine, if belated shaman. He was one of the last people who could handle a canoe, a rifle, a lost language, or a kinship system and do them all superlatively well. He also knew the lore of ferns and how to catch lizards."

"But you yourself said he was a shaman," protested the administrator.

"Look," I said, "he believed certain things, practices, if you will, that the hunters know. Before he died and before he gave up hope, there was a tribe, the Seneca, who brought him to their reservation for a healing ceremony. Sick as he was, he went, you know. It tells you something. It tells you also how they felt about him. His office was the last Hudson's Bay post south of the Canadian border. A lot of odd things and odder people had a way of turning up there. Including myself." I grinned, to take the sting off the words, but I meant them. "You won't find the likes of us any more," I added. "We aren't relevant. Frank would have died of frustration if he had had to face the students of the sixties.

"It was different then." I gestured hopelessly. "None of us was relevant, or wanted to be, or expected to be. Frank's office was a refuge for misfits who became scholars because of him. Maybe it wasn't the teaching exactly. That's where the shaman bit came in. He believed in something, perhaps the animal powers the Indians invoked in the shaking tent rite. Animals spoke from it; they don't speak to us any more. I never knew quite what he believed, and it troubles me now. I might be a better man if I knew. No, he wasn't a shaman, he was the last of wild America. It's gone now, gone. I am a stranger in my own department."

"You always were, Loren," said my superior cryptically. "Like a man come in with furs to warm himself at the stove, but not to stay. That's what Frank knew, that's your secret."

I didn't answer. He strode off confidently down the walk.

"Well," I thought, but I did not follow it up. It wasn't the time to ask for Captain Riley. When the real trouble comes you ask for no one. There isn't time. Still, why had his wild son Tim held that single seat in the coach for me, the stranger? I would never know now. I had stayed behind.

CHAPTER 13

Madeline

MADDY was what we called her familiarly. Madeline was her real name and she was a prima donna and a cat—in that order. Maddy was a cat that bowed, the only one I have ever encountered. She is part of my story, what one might call the elocution or stage part. We patronized each other. Maddy performed her act, and I assumed the role of her most ardent admirer. In discharging this duty I learned a great deal from Maddy, my patroness, whom I here acknowledge.

I have known a good many cats in my time—some that scratched, some that bit, some who purred, and even one who, by my standards at least, talked. I liked Maddy, I suppose, because we had so much in common. Maddy was an isolate. Maddy lived with three other more aggressive and talented animals who took the major attention of my host. Maddy, by contrast, was not so much antisocial as shy, when you came to know her. She wandered a little forlornly in back rooms and concealed herself under furniture, or in a recess above the fireplace.

Maddy, in short, wanted a small place in the sun which the world refused to grant her. She was at heart simply a good-natured ginger cat in a world so full of cats with purrs less hoarse that Maddy, like many of us, had learned to slink ob-

scurely along the wall and hope that she might occasionally receive a condescending pat. Nothing was ever going to go quite
right for Maddy. In this we reckoned without poor Maddy's
desperation. She discovered a talent, and I, at least, among her
human friends, was appalled at how easily this talent might
have gone unguessed, except for a chance episode and an equally
uncanny tenacity on Maddy's part. Maddy learned to bow.

Perhaps you may think, as a human being, that this is a very
small accomplishment indeed. Let me assure you that it is not.
On four feet it is a hard thing to do and, in addition, the cat
mind is rarely reconciled to such postures. No one among us in
that house, I think now, realized the depths of Maddy's need
or her perception.

It happened, as most things happen, by accident, but the
accident was destined to entrap both Maddy and ourselves.
She was ensconced in her favorite recess upon the mantle of the
fireplace, watching us, as usual, but being unwatched because
of the clever gyrations of one of her kindred down on the floor.
At this point, Mr. Fleet, our host, happened to stoop over to
adjust a burning log in the fireplace.

Easing his back a moment later, he stood up by the mantle
and poked a friendly finger at Maddy, who came out to peer
down at the sparks. She also received an unexpected pat from
Mr. Fleet. Whether by design or not, the combination of sparks
and the hand impinging upon her head at the same time caused
Maddy to execute a curious little head movement like a bow.
It resembled, I can only say, a curtsey, an Old World gesture
out of another time at the Sun King's court.

Maddy both hunched her forefeet and dropped her head. All
she needed to complete the bow was a bonnet or a ribbon.
Everyone who saw applauded in astonishment and for a few
moments Maddy, for once, was the center of attention. In due
course that would have been the end of the matter, but a
severe snowstorm descended over our part of the state. We
were thus all housebound and bored for several days. This is

where Maddy's persistence and physical memory paid off; on the next night upon the mantle she came out of her own volition, and bowed once more with precisely similar steps. Again everyone applauded. Never had Madeline received such a burst of affectionate encouragement. If there were any catcalls they could only have come from her kin beneath the davenport. Her audience was with her. Maddy seized her opportunity. She bowed three times to uproarious applause. She had become the leading character in the house. The event had become memorable. Maddy, no more than a dancer, would forget the steps and the graceful little nod of the head. It became an evening routine.

I have said that in the end both Maddy and her audience were entrapped. It happened in this way: finally the snow went away and we, all except Maddy, tried to resume our usual nocturnal habits—the corner bar, the club, the book. But the bow had become Maddy's life. She lived for it; one could not let her down, humiliate her, relegate her unfeelingly to her former existence. Maddy's fame, her ego, had to be sustained at any cost, even if, at times, her audience was reduced to one. That one carried, at such times, the honor of the house. Maddy's bow must be applauded. Maddy would be stricken if her act began to pall.

To me the act never did pall. More than once I gave up other things to serve as a substitute audience. For, you see, I had come to realize even then that Maddy and myself were precisely alike; we had learned to bow in order to be loved for our graceless selves. The only difference was that as a human being living in a more complex world it had taken me longer to develop the steps and the routine. I talked for a living. But to talk for a living, one must, like Maddy, receive more applause than opprobrium. One must learn certain steps.

I was born and grew up with no burning desire to teach. Sea captain, explorer, jungle adventurer—all these, in my childhood books, had been extolled to me. Unfortunately my reading had not included the great educators. Thus upon completing my doctorate I had no real hope in those still depressed times of 1937 of finding a university post. While I was casting about for a job among newspaper folk of my acquaintance, word came that I had been proffered a position at the University of Kansas.

Most of the midwestern universities of that period had joint departments of sociology and anthropology, or at best one tame anthropologist who was expected to teach in both fields. When I appeared that fall before my first class in introductory sociology I realized two things as I walked through the door. I did not dare sit down. I did not dare use my notes for anything but a security blanket to toss confidently on the table like a true professor. The class was very large. A sizable portion of the football squad was scattered in the back row. I was, I repeat, an isolate like Maddy. If I ever lost that audience there would be chaos. The class met every day in the week.

Each night I studied beyond midnight and wrote outlines that I rarely followed. I paced restlessly before the class, in which even the campus dogs were welcome so long as they nodded their heads sagely in approval. In a few weeks I began to feel like the proverbial Russian fleeing in a sleigh across the steppes before a wolf pack. I am sure that Carroll Clark, my good-natured chairman, realized that a highly unorthodox brand of sociology was being dispensed in his domain, but he held his peace. By then everything from anecdotes of fossil hunting to observations upon Victorian Darwinism were being hurled headlong from the rear of the sleigh. The last object to go would be myself. Fortunately for me, the end of the semester came just in time.

At the close of the first year I had acquired, like Madeline the ginger cat, some followers. I had learned figuratively to bow and I was destined to keep right on bowing through the next thirty years. There was no escape. Maddy had taught me how necessary it was that one's psyche be sustained. An actor, and this means no reflection upon teaching, has to have at least a few adoring followers. Otherwise he will begin to doubt himself and shrink inward, or take to muttering over outworn notes. This is particularly true in the case of a cat who has literally come out of nowhere to bow under everyone's gaze on a fireplace. Similarly I had emerged as a rather shy, introverted lad, to exhort others from a platform. Dear Maddy, I know all you suffered and I wish I could think you are still bowing to applause. You triumphed over your past in one great appreciative flash. For me it has been a lifelong battle with anxiety.

This anxiety reached its climax in the 1960s. I am not a political speaker but for some obscure reason I was occasionally to find myself on violent campuses delivering lectures on such subjects as "Ice, Time, and Human Destiny" just at a period when some new bewhiskered idol was delivering a competitive address on smut, or extolling the virtues of fire-bombing the whole place into oblivion. I have never quite understood why my recondite subject matter drew the audiences that it did, unless the campuses were so hot that, as in the case of the title just quoted, even students wanted a breath of cool air off the ice sheet. I cannot think otherwise in one instance because shortly after I departed a building was blown up and a death resulted.

Then in the months that followed I received an invitation to lecture in Halifax, Nova Scotia. I turned the matter over in my mind. Louis Agassiz, the emigrant Swiss scholar whose name is as deeply associated with the ice age as Darwin's is with evolution, had come ashore there in 1846. His first act had been to journey into the hills behind the town to see if he could observe glacial striations in the stones of the New World. He did indeed find them, and the glacial hypothesis moved definitively to America. Historically this interested me. I was simple enough to think of Canada as less prone to violence than the United States. I was assembling a collection of my essays at the time and

I chose one of a biological nature. I looked forward to a trip happily remote from the turbulent world through which I had been journeying in the States.

In Halifax I was received pleasantly by my faculty hosts and shown about at length. Shortly before I was to enter the auditorium, a brother professor drew me aside. "I should warn you," he said. "A group of Maoist students may make trouble. They will circulate a leaflet in the aisles."

"Oh?" was all I could venture. I did not see then, nor do I now, how my talk, as the leaflet recorded, had any connection with the fact that I was an American "imperialist" who enjoyed "romping through sunflower forests and snowy woods." Until I came to Halifax up there in the subarctic, these pleasures of mine had seemed harmless enough. Now they had come under the suspicious eye of the proponents of the People's Republics of China and Albania. They were "anti-people" pleasures. The one heartening thing in the leaflet to be distributed was that it referred to "Eiseley and his cohorts." I had never realized before, being of a rather solitary cast of mind, that I had cohorts. Wherever they were, they were not massed at my back as I received some ironic best wishes and stepped to the podium.

I gave my address. No one interrupted. Indeed one scarcely could have, because, while not romping in the woods this time, my thoughts were rather far afield from the regions of Marxian dialectic. After I had concluded, some friendly people came forward to have books autographed and to chat. The audience drained away; the Maoists were silent. Considerably later, so much so that my departure must have been deliberately awaited, I broke away from my little group of well-wishers and, led by one of my host colleagues, walked up the aisle to the door. A tall youth lounging there in the shadows sidled up and spoke in my right ear. I did not hear him distinctly, but sensing his unfriendliness I shrugged and passed on.

"You didn't hear him?" questioned my host.

"No," I said.

"He called you a name, a very ugly name."

Indifferently I started to say, "no matter," when the import of the thing struck me in all its diabolical cleverness. The youth had actually tried to provoke a fight and, thinking back upon the long way I had come, he might well have succeeded except that he had muttered in the wrong ear. The next morning there would have been a headline in the paper. I could see it in block-letters, "AMERICAN PROFESSOR ASSAULTS CANADIAN STUDENT." Typical of these aggressive fascists from south of the border, "the heartland of the beast," as the Maoists had aptly phrased it in their leaflet. Perhaps they had had a photographer lurking in the background.

I hold an honorary doctorate from a Canadian university and I shall not commit the ancient fallacy of categorizing a nation by a few individuals, but I am glad that Madeline never had to face such people as I had faced in Halifax, nor have her little polite bow be taken for the machinations of an "arch criminal," as mine was by planned intent.

I will conclude my account of adventures on the speaker's platform with my travels to a midwestern university where I was warmly greeted upon commencement day. As it turned out, I departed bruised in every part of my body, yet no one had intended it so. Chance produced what the Maoists would have delighted in achieving.

The trip began inopportunely in New York. Rushing on a short schedule from one engagement to another, I was held up in traffic and just barely reached my plane at Kennedy airport before departure time. I hurled myself aboard after a long run down a corridor, only to find that very shortly after I boarded there was an announced delay caused by mechanical difficulties. The delay stretched into three hours. This meant that I would arrive at my midwestern destination at midnight. From the airport I was to be driven for another hour into the countryside.

The people assigned to collect me had waited patiently. We drove off into the night. Utterly weary from the flight and the tedious delays, as well as my long sprint to catch the supposedly departing plane, I was finally brought into the lobby of a run-down hotel. "The school has just bought the place," explained the man who had chauffeured me, a little apologetically. "Here is your key." It was big and looked to be of Civil War vintage.

"We'll meet you here at eight in the morning for breakfast," my companion added as I studied the key a little blearily. "The commencement is outdoors."

"All right," I said, and stumbled away to the elevator. My room had a slowly revolving window fan that must have been as weary as I. I dropped my clothes on a chair and climbed into bed. That was the last I knew until the operator summoned me at seven.

Now to understand my morning ordeal it is necessary to explain the nature of the room I occupied. It was entered by a short hallway. On one side of the hall was the entrance to the bathroom, elevated a good half foot above the level of the floor. A rusty single pipe shower curled over a bathtub with a V-shaped bottom, holding no rubberized mat. Directly across the hall was an open closet full of dangling coat hangers. I should have known it was a booby trap, but I was still groggy from the interminable delays of the night before. I stripped and climbed into the tub.

The shower descended in a feeble stream while I lathered myself. Then the inevitable happened. I slipped on the V-shaped bottom of the confined tub. Only one thing prevented me from cracking my skull. In an instinctive desperate lunge as I started to fall, my elbow had caught, through the shower curtain, in the antique marble washbasin beside the tub. Desperately I heaved myself out of the tub. My elbow ached and was bruised but not broken. The rubber curtain had partly shielded it. Shocked by my fall I decided to towel off on firmer footing. I stepped to the bathroom door, forgetting as I did so the raised step.

The next moment I was lying naked in the coat closet across the hall while coat hangers showered down upon me. I looked up in stunned astonishment. I thought of Madeline. The show must go on. I crawled out upon the rug. More bruises, but miraculously no breaks. I dressed painfully and descended to the coffee shop. My beaming hosts arose.

I have always since that day detested open-air commencements. One's speech, fortunately for me on this occasion, tends to rise and be lost in the spring air. Parents rush about frantically beneath one's pedestal, snapping pictures of their young immortals. Family dogs frisk in the aisles, unimpressed by words. Also, in that time, there were the hard-core youngsters who thought the world was going to fall by the mere weight of their clenched right fists brandished as they marched past the platform, or those who deliberately avoided academic garb or, worse, wore fragments of it. I groaned from both mental and physical bruises. It's time we stopped, I sighed under my breath to the lost Madeline, my teacher, but I knew Madeline would never have stopped, nor would I. Something got into your blood, like the hunger of an old vaudeville trooper. Maybe it was a remaining wanderlust in me. There was polite applause. At last the ceremony was over.

"Come," said the president. "We are lunching with the trustees." I limped inconspicuously after him. There was a raised, temporary dais at the end of the dining hall. Curtains were draped behind it. The president led me gravely to the curtains, swept them aside and pointed downwards. I saw an array of hard-edged two-by-fours, but no planking back of the curtains. "I just thought I should point out to you" said the president blandly, "that the flooring does not extend beyond this point."

"Of course," I said mechanically. I had no intention of wandering. I wanted never to get out of a chair again. The luncheon passed quietly. We arose in unison. The affair was breaking up.

Before I could get off the platform one of those formidable dowagers who are patronesses of the arts advanced upon me with a barrage of questions.

"How did I write? Under what circumstances? What was my inspiration?"

I shifted from one foot to another. A police interrogation would have been preferable. She leaned forward. Like any terrified male I stepped backward. The president was momentarily engaged. The formidable woman once more moved relentlessly upon me. Again I executed a backward step.

For the second time that day the world unexpectedly shifted. I was lying on my back among dusty timbers. I had vanished. So, instantaneously, had the dowager. I will never know whether she had actually existed or was an emissary from beyond the pale. Somewhere above me the president was gazing down with mild solicitude. "Are you hurt?" he enquired.

"My God, that yellow-livered ———" I started to bawl to the assembled guests, but I didn't. I thought of Maddy. The show had to go on. Like Maddy, whose furry head had dropped lower at every performance. I arose from the timbers that might have snapped my neck. I bowed to the guests, who probably, one and all, thought I was blind drunk.

"I'm fine," I said. "I do this—well, not every day but almost." This last I whispered under my breath.

"Those timbers, we must change them next year," murmured my host. "You're not the first, you know. Our warnings"— here he sighed—"are not always heeded. Come now and see the stained glass in our chapel."

Surreptitiously I tested my bones. Once more, like Maddy, I made a little step and bowed. "I will be delighted," I said. Maddy, if she could have talked, would have said that, too. Good old Maddy, she knew what to do. To bow. It was our final, constant necessity, Maddy's and mine.

Some years later I was asked to repeat my performance at the same university. Regretfully I wrote that I had another engagement. I could still bow but they did not know that my limbs were now more fragile. In fact, they did not know what they were asking, but I did. I wonder if they still use the trick of the curtain, the beams, and the dowager. I knew if I went there again I would never escape alive.

CHAPTER 14

A Small Death

I HAVE remarked that I was born in the central plains, compacted out of glacial dust and winter cold. I see animal faces as readily as though I sat with my mother's one blighted gift in a Cro-Magnon cave. The religious forms of the present leave me unmoved. My eye is round, open, and undomesticated as an owl's in a primeval forest—a world that for me has never truly departed. The boy who watched the gold crosses of his childhood swept up by a man with a scythe would scarcely hope too much for what lay forward.

In one of those golden Octobers that fell between the wars, I made a journey somewhere in southern Kansas. The friend who went with me was Claude Hibbard, a paleontologist who died only a year or so ago after an intense and devoted career given to the life of the past. Claude had got word of a farmer who had found an enormous long-horned bison skull in a gravel pit. The man had it hanging out in the weather under the eaves of his barn.

The skull proved to be a beautiful example of one of the least-known bisons, the horns having the sweep of one's extended arms. The names change in the taxonomists' reports but the size of the great beast never; he belongs to the Middle Ice. The farmer coveted the skull as a barn decoration, little realizing that it would slowly flake away in the winter rain and snow. It was our errand to try and secure the bones for the museum of the university.

We found our man in the field. Claude bore the brunt of the negotiations, for he was not only a native Kansan himself but came of a farming family. The farmer was reluctant. He was harvesting corn and we followed the wagon, automatically tossing up the ears as the talk progressed in slow country fashion. We went up one row and down another, serving as an unpaid labor force.

Claude knew what he was doing. I slung in ears of corn and an occasional word. So did Claude, but his remarks were all of crops and farms and weather. The golden afternoon waned to its close, a blue frost was rising. "That skull," said Claude finally, " 'tisn't right it should be wasted here. If you gave it to the Museum you could have your name on it, 'donated by,' or 'on permanent loan.' Here it'll just go to pieces. It needs to be treated. It's been a long time in that gravel bank." It had, too. The gravel bed in which the skull had been miraculously preserved had been deposited long before the last great ice advance.

"Well," said the weatherbeaten farmer, "I dunno." We threw the ears of another row into the high-sided wagon while he meditated reluctantly. "I reckon," he said, "if my name—"

"It'll be in the books," said Claude. "Let's just get this last corn row. Might as well before dark."

So that was the way it ended, but not for me. The curtain of time had lifted an instant in that slow smoky autumn. I could almost see the great herds from which this giant came. Across them swirled the blizzards of my childhood in which I used to go out alone because no one could find me thirty feet away.

As we left, the leaves in the wood were red and coming down. It was, I think, the last time I saw Claude, though his professional papers came faithfully to me for many years. He was trying, no doubt, his own keys to the past. As for me, I have come to think I am moving in an endless extension of that single

Kansas autumn. I am treading deeper and deeper into leaves and silence. I see more faces watching, non-human faces. Ironically, I who profess no religion find the whole of my life a religious pilgrimage. The origins of this hunger are as mysterious as the reasons why we, who are last year's dust and rain, have risen from that dust to look about with the devised crystal of a raindrop before we subside once more into snow and whirling vapor. But, however that one autumn may still color my memory, life is complex; it changes, and my world was destined to change with it.

The change was mixed with many things in my life—a growing disillusionment with some aspects of scientific values, personal problems, abrasive administrators, humanity itself. In short, the war had finally come to Kansas and transformed the pleasant, sleepy little town of Lawrence overnight.

When the first Polish towns were burning, some of my students had laughed, far on there in the isolationist mid-country, and called the newsreels "propaganda," in the disillusionment that still lingered after the first World War. Then came Pearl Harbor. A giant powder factory arose in the environs of Lawrence and hundreds of people poured in from the backwoods of the Ozarks and Oklahoma. Almost overnight Lawrence became a boom town in which prices rose, people began to distrust each other, and desperate workers trudged the streets seeking shelter. From wary isolationism a few students turned to militant patriotism in a way sometimes reminiscent of the excesses of the first World War. As was true of most universities, training programs for enlisted men under discipline soon emerged.

My unmarried friends quickly disappeared, either called up in the reserves or in the draft. As a physical anthropologist by training, with a background in anatomy and biology, I was shifted as essential into the pre-medical program. We taught enlisted young reservists almost around the clock. Summer school was no longer a matter of choice, and I labored on through attacks of violent asthmatic hay fever which afflicted me in that climate. I had to refurbish forgotten knowledge, and my fellow anatomists, including Dr. Henry Tracy, the grey wise head of the department, could not have been kinder.

My troubles, however, were endless. I was the last of a line that had volunteered and fought in almost every war since the War for Independence. Yet here I was confronted with an utterly impossible situation. My mother, I had come to know, was committable without the care and attention of my aunt. All this had been certified and was on record. Both women were totally dependent upon my support. Moreover, my wife came home one day from the doctor's office to report a diagnosed illness which necessitated surgery. We had to borrow money. Still I fretted, although by now four million men were under arms and the landing in Europe was drawing near. Perhaps it was foolish of me, in retrospect, but I was still young and there was a family tradition. I wanted to go. This impulse was to be suddenly augmented.

One day I learned, in some way now forgotten, that there was a need for men capable of staffing military government in the islands of the Pacific being slowly overrun by the island-hopping technique of the Pacific war. Such an assignment would have solved the salary problem. My training and teaching in social anthropology seemed to offer some hope.

I was given a physical. My eyes were not up to the military standard for a professional officer. I argued a little with the doctor. "This is governmental activity," I protested. "There is a supposed vital need. I can see enough with glasses. I've been doing all that this job would require for almost six years. If a belated shell comes along, what difference will glasses or non-glasses make?"

"Now the ears," he said imperturbably. I sat in a makeshift booth and heard, with my left ear, the faint ticking of a watch held some distance away. Then my right ear was tested. Suddenly I realized that, though it was not badly off, I was not sure of the ticking. Something was slightly wrong. I thought of the forgotten episode of my first year in graduate school. The other routine procedures followed.

In a week or so I received my rejection. I have often wondered in the years since, as I have come to know a little more about military intelligence, whether it was actually the minor physical defects or the existing affidavits about my mother that made the difference. Later, I was to know men who seemed never to have encountered the troubles that beset me. The university was, of course, struggling to maintain its staff.

In elaborating this background, however, I have neglected a dog whose plight actually affected me more than the turmoil that swept around me. It was a small death in that war now long since done. I do not know why I remember it with such pain, but yes, yes I do. I remember it with an uncertain guilt, just as I remember my last glimpse of the desperately running mongrel beside the train in the cruel days of the Depression. As it chanced, I was assisting one of my medical superiors in a cadaver dissection. He was a kind and able teacher, but a researcher hardened to the bitter necessities of his profession. He took the notion that a living demonstration of the venous flow through certain of the abdominal veins would be desirable. "Come with me to the animal house," he said. "We'll get a dog for the purpose." I followed him reluctantly.

We entered. My colleague was humane. He carried a hypodermic, but whatever dog he selected would be dead in an hour. Now dogs kept penned together, I rapidly began to see, were like men in a concentration camp, who one after the other see that something unspeakable is going to happen to them. As we entered this place of doleful barks and howlings, a brisk-footed, intelligent-looking mongrel of big terrier affinities began to trot rapidly about. I stood white-gowned in the background trying to be professional, while my stomach twisted.

A SMALL DEATH

My medical friend (and he was and is my friend and is infinitely kind to patients) cornered the dog. The dog, judging from his restless reactions, had seen all this happen before. Perhaps because I stood in the background, perhaps because in some intuitive way he read my eyes, perhaps—oh, who knows what goes on among the miserable of the world?—he started to approach me. At that instant my associate seized him. The hypodermic shot home. A few more paces and it was over. The dog staggered, dropped, and was asleep. The dose was kindly intended to be a lethal one. He would be totally unconscious throughout the demonstration. He would never wake again. We carried him away to the dissecting room. My professional friend performed his task. A few, a very few, out of that large class, crowded around closely enough to see.

The light was pushing toward evening. The dog was going; this had been his last day. He was gone. The medical students attended to their cadavers and filed out. I still stood by the window trying to see the last sun for him. I had been commanded. I knew that, even if I had not been in the animal house, the same thing would have happened that day or another. But he had looked at me with that unutterable expression. "I do not know why I am here. Save me. I have seen other dogs fall and be carried away. Why do you do this? Why?"

He did not struggle, he did not bite, even when seized. Man was a god. It had been bred into this creature's bones never to harm the gods. They were immortal and when they touched one kindly it was an ecstasy whose creation their generations had never understood, because for them there was only one single generation, their own.

I mentioned my feelings once, years later, to a friendly physician. I was a scientist. I was groping for some way to explain. "My friend," he said, "this is necessary. You are imagining things. Dogs don't think like that. You merely thought he was looking at you. You are not in medicine. You do not know the necessity of these things." He, too, was a good man, but I remembered persistently the indifferent class that had gained little from that experiment. I would venture that not one of them remembers it now in his gleaming office. And the dog might have had just one more day, one more day, even in the animal house. One more day of life, of sentience.

I shook my head wearily. There is a man, a very great experimentalist, who has said that to extend ethics to animals is unutterable folly. Man cannot do this and learn, learn even to save himself. Each one of us alive has inevitably, unknowingly taken something from other lives. I thought of my steeled professional friend. I thought, though I did not say it to my concerned acquaintance, that the experiment I had witnessed in my judgment was needless.

Just one more day, those beseeching eyes continued to haunt me. They do still. I have stood since in some of the cleanest, most hygienic laboratories in the world. I have also watched dirty, homeless dogs or cats trot on to what must have been for most of them starvation, disease, or death by accident. I have never called a humane society because I, too, am an ex-wanderer who would have begged for one more hour of light, however dismal. Rarely among those many thousands have I been able to protect, save, or help. This day I have recounted is gone from the minds of everyone. As for me, I have sought refuge in the depersonalized bones of past eras on the watersheds of the world.

The beachheads were finally established. The Germans' last desperate offensive toward Bastogne and the Channel ports was contained. Japan's sea empire was tottering. Men of my age and condition were freer to move again. I was proffered an administrative post and a full professorship at an Ohio college under a dean who was both a fine historian and a great man.

I spoke to Henry Tracy, the chairman in anatomy. "Eiseley,"

he said wistfully, "would you like to be a physician? There are those here who could arrange admittance to Med School for you." I was deeply moved.

"H'ie," I said, using his affectionate nickname, "I wish I might have heard those words ten years ago. But I'm thirty-seven; I have just been offered a very good post in Ohio. I have the same dependency problems. My wife must still have care. It's too late, but I am prouder of what you have just said than anything that will ever come to me."

"You will need some money to get there," he considered.

"Yes, H'ie," I said, "not much, but a little. I'll give you a paper for it. And you will come to see us. I must thank the surgeon in Kansas City as well. He gave me professional courtesy; he wouldn't have had to do it. Someone must have told him."

"Well," said Henry Tracy and stopped. He was the only man I ever knew who could see a body, even a walking body, as a three-dimensional collection of pipes, pumps, and pulleys. He would have made a fantastic plumber. Just to complicate things, he had begun his academic career as a classicist. Maybe his interior view of people had forced the change in his profession.

I shook hands with the few men with whom I had started my career. Lawrence would never go back to what it had been. The administrative staff I knew is long since gone.

Before I left Lawrence, the old anatomy building burned in the night. I was routed out of bed by an official, who told me to get up on the hill immediately and see to the bodies floating in the tanks below the floor. "You are confusing me," I said, "with another man." So, as in the hatchery days, the bridges literally burned behind me.

Years later, at a public function in another city, I stood by accident just behind one of the administrators of that day. I turned and went quietly out into the night. About what could we have conversed? The laughter, the ugly laughter under spotlights at the spectacle of naked cadavers being piled into trucks out of the smoking ruins of a building? For the students, crude as their behavior was, it was an attempt at total evasion. "I shall never be as these are," they had tried to reassure themselves with raucous laughter. "Death is a joke. We, we, are the immortals, the golden boys and girls." Have they learned differently in the years that followed? I lift my grizzled head. They are close now upon my heels. I can hear them. I can hear the night frost split stones in deserts. Men are softer than stone, much softer.

A number of years ago, across Walnut Street from an office I once occupied, there used to be a series of dilapidated row houses occupied by elderly people, mostly men, who lived upon welfare or social security checks. Some of them were winos or city derelicts, who, on sunny days, drowsed upon university benches. It is true that they were not good to look upon and sometimes one of them would sprawl helplessly upon the sidewalk, a bottle still clutched in his hand. Sometimes what my grandmother would call the "dead wagon" would come and take one away.

The city, in time, condemned these properties so that the University might construct a dormitory on the spot. An ugly district police station that had once maintained a kind of order in that neighborhood disappeared. So did a street, the derelicts, and the houses.

Before all was quite leveled, something strange happened. A few abandoned old dogs refused to go. They were lying, in a sort of momentary local return to the stone age, behind building blocks and in depressions that sheltered them from the bitter weather. They retained instincts from a past older than ours. They had accepted desertion, since they had never been well cared for; they had accepted without question the destruction of all they had once lived amidst. They sat like wolves in the wreckage, nosed about, or slept. Probably the city would gather them up, and perhaps, if I walked in a certain direction I deliberately never took, I would eventually hear them barking for release in the animal rooms by the laboratories.

I crossed the street and came to them among the stones. Maybe, I thought, just maybe, if we pass, under their thick wild hides they may preserve for a time a dim memory of a visiting god who could not save himself but whose touch wrought something ineffable. From among the stones an old brown derelict crept forward and ran a wistful stroking tongue across my hand. I knelt and spoke to him gravely. Perhaps, after all, the little we knew of love may linger a few seasons in the wild pack that roams the final rubble of the cities. For a century or two the pack may lift its ears to a rockfall or sniff with lifted hair at a rain-worn garment that touches an old racial memory and sets tails to wagging expectantly. Some dim hand that they all feel but have never known will pass away imperceptibly. And when that influence is no longer felt nor remembered, then man will in truth be gone.

Still, crouched on my knees in the dust and white rock of this field where the homeless dogs lay tail to wind, I thought perhaps it might not happen, that perhaps the lightning would only seek out the most of men, and that this old brown wolf and I would lie beside yet another ruin and watch the stars come out through the bent ribs of skyscrapers.

"If you would come out of your doors and your stonework," the patient stroking tongue tried to persuade me, "we could lie here in the dust and be safe, as it was in the beginning when you, the gods, lived close to us and we came in to you around the fire."

I stroked his head gently so that he might remember me, and walked toward the station. He lifted his ears. He did not understand the gods nor why they persisted in going so far away. I felt a little lonelier from the touch of his rough tongue. Men, too, it seems, have a bit of common dog in their natures. But in the shelter by the stones the dogs slept and thought I would be coming back. They have an enormous, unquenchable, betrayed trust in man. I think they will still be waiting when the first wild oak bursts through the asphalt of Market Street.

CHAPTER 15

Willy

NOWADAYS, when I pass along the walk where the tenements were leveled, I persist in seeing, not the massive architecture of the new building, but the dust and the stones where, in my mind, the dogs still lie in the October wind. Man is a strange creature. I look upon this great building with its inner fountains and amenities, and though it is well over ten years since it was constructed, I see right through it to the bare field left by the demolition of the slum. Something has seized and held me there, created what is even more real than what currently exists. Perhaps it is my mother's unrestrained clairvoyant eye. I cannot control it.

My sight comes and goes of its own volition. Just today, for example, I turned a corner and passed a girl whose name I almost spoke. She had a certain cast of cheekbone and a merry eye. Then, with a wrench, I realized I was seeing someone in youth who—if she still lives—is in her sixties. But here was the face, and I had immediately reached backward into time beyond the elapsed years. I had to restrain myself from speaking. We go away and the other person stays eternally young, to be seen at rare and sudden intervals on a far street corner, or down a pathway in the park. Time never touches such people. It is we who, in the very moment of speaking, draw back in embarrassment. We are never recognized; we have grown old.

In all the questioning about what makes a writer, and especially perhaps the personal essayist, I have seen little reference to this fact; namely, that the brain has become a kind of unseen artist's loft. There are pictures that hang askew, pictures with outlines barely chalked in, pictures torn, pictures the artist has striven unsuccessfully to erase, pictures that only emerge and glow in a certain light. They have all been teleported, stolen, as it were, out of time. They represent no longer the sequential flow of ordinary memory. They can be pulled about on easels, examined within the mind itself. The act is not one of total recall like that of the professional mnemonist. Rather it is the use of things extracted from their context in such a way that they have become the unique possession of a single life. The writer sees back to these transports alone, bare, perhaps few in number, but endowed with a symbolic life. He cannot obliterate them. He can only drag them about, magnify or reduce them as his artistic sense dictates, or juxtapose them in order to enhance a pattern. One thing he cannot do. He cannot destroy what will not be destroyed; he cannot determine in advance what will enter his mind.

By way of example, I cannot explain why, out of many forgotten childhood episodes, my mind should retain as bright as yesterday the peculiar actions of a redheaded woodpecker. I must have been about six years old, and in the alley behind our house I had found the bird lying beneath a telephone pole. Looking back, I can only assume that he had received in some manner a stunning but not fatal shock of electricity. Coming upon him, seemingly dead but uninjured, I had carried him back to our porch and stretched him out to admire his color.

In a few moments, much to my surprise, he twitched and jerked upright. Then in a series of quick hops he reached the corner of the house and began to ascend in true woodpecker fashion—a hitch of the grasping feet, the bracing of the tail, and then, wonder of wonders, the knock, knock, knock, of the questing beak against our house. He was taking up life where it had momentarily left him, somewhere on the telephone pole. When he reached the top of the porch he flew away.

So there the picture lies. Even the coarse-grained wood of the porch comes back to me. If anyone were to ask me what else happened in that spring of 1913 I would stare blindly and be unable to answer with surety. But, as I have remarked, somewhere amidst the obscure lumber loft of my head that persistent hammering still recurs. Did it stay because it was my first glimpse of unconsciousness, resurrection, and time lapse presented in bright color? I do not know. I have never chanced to meet another adult who has a childhood woodpecker almost audibly rapping in his skull.

Robert Louis Stevenson, who had an eye for such matters, maintained that there are landscapes that cry out for a story and I suspect that his tale of *The Merry Men,* along a tide-ripped coast, revolves about some personal vision of his own. Similarly Dickens once spoke of the "cold wet shelterless streets of London," something I myself can attest to from memories of one foggy night in those same streets.

Amongst this odd collection of pictures I must confess that much of historical importance has passed me by. I do not travel to political rallies and it has not been my fortune to be present at the scene of great events. On the whole, as I pause to examine this lost studio in my head, the animals outnumber by far the famous people I have met. If I sense a dearth of presidents, I have still encountered, though he looked right through me, one magnificent snow leopard, and I have also danced with an African crane. The crane, which is nearly as tall as a man, has an intricate mating dance. I was once strolling in the Philadelphia zoo when I came upon one of these birds solitary in a barely retaining enclosure.

In the animal world lines of definition are not as severely drawn as in the civilized one that we inhabit. This bird, acting under the impulse of spring, made some intricate little steps in my direction and extended its wings. Now I too believe in friendliness and spring festivities. I realized that the bird saw me as a vertical creature of the proper appearance to be a potential mate. To simplify things for her unlettered offspring, nature imparts, as in this case, a recognition of the vertical. After all, what is a face to a creature with a large bill? But then, unfortunately, in order to prevent, in her wisdom, unwise mixtures such as I and this crane potentially represented, nature insists upon an extremely complicated recognition dance. If one fails the steps and gestures, nothing is going to happen.

I fitted the vertical line pattern all right and I tried to be a good sport about the rest. I extended my arms, fluttered and flapped them. After looking carefully up and down the walk to verify that we were alone, I executed what I hoped was the proper enticing shuffle and jigged about in a circle. So did my partner. We did this a couple of times with mounting enthusiasm when I happened to see a park policeman sauntering in our direction. I dropped my arms and came to a direct, meditative halt.

The bird, too, paused uncertainly. There were now two attractive vertical figures, but they really did not seem to know the approved steps. Furthermore, not having read up on African cranes, I was a bit uncertain about the sex role I was playing. Male, female? I looked at the policeman. He looked at me. Suddenly I felt it best to leave the vicinity. Three is a crowd at moments like this. I walked away with careful unconcern in the direction of the small mammal house.

But why should my dance with a crane supersede in vividness years of graduate study? One can see a certain lack of disciplined control in a mind of this sort. Either that or the artist eye of my deprived mother lingered in me so that I was too much taken with color and form. I remember the vast wastes of the Mohave but, much more than that, I recall a baby ground squirrel that I came suddenly upon sunning himself in some fresh-turned earth beside the family burrow. His mother must have been careless, for here was her little waif blissfully lying on his back and patting his stomach. He looked up at me without a trace of fear as I stood over him. There the image stays, yet close to fifty years have passed: one ground squirrel patting his paunch.

I think, you know, it is the innocence. A violent dog-eat-dog world, a murderous world, but one in which the very young are truly innocent. I am always amazed at this aspect of creation, the small Eden that does not last, but recurs with the young of every generation. I can remember when I was just as innocent as that baby ground squirrel and expected good from everyone, as a puppy might. We lose our innocence inevitably, but isn't there some kind of message in this innocence, some hint of a world beyond this fallen one some place where everything was otherwise? Why else do infants peer up with humorous, arch visages, as when I briefly scratched the ground squirrel's belly and saw him wriggle?

These are the pictures that haunt my mind so that I stare through brick or stone as if it were not quite there. I know that I have written of harsh events and of those memories I wish that some might be effaced, scratched over with great black erasure marks, but this is not the way the essayist writes. He sees as his own eye dictates. Once, far north in Canada, I came upon a tremendous pile of boulders tossed about like houses in a hurricane. I was dwarfed beside them. They were remnants dropped from the retreating Labradorean ice cap. The huge stones were spilled as carelessly as a child's playthings upon a sidewalk. They could be picked up again as readily, the timeless eye looking through the boulders predicted, just as the little blue lakes between the stones could once more be frozen permanently and rise to obliterate the present countryside. There is a place like this as far south as Nebraska, where in the evening light everything shifts and changes and the transported granite takes on the shapes of marching mammoth.

These pictures reduce us to miniscule proportions, but I have so long wandered among eroded pinnacles and teetering tablestones that I have felt as lost as an insect drifting into a colossal ruin, not alone of earth, but of ages. I suppose if these stones had been glimpsed by Stevenson he would have said that they demanded a story, a human story of equal proportions. Being of a different vision, I can only say the story already lies there.

Nothing human will compass it save this: we, mankind, arose amidst the wandering of the ice and marched with it. We are in some sense shaped by it, as it has shaped the stones. Perhaps our very fondness for the building of stone alignments, dolmens, and pyramids reveals unconsciously an ancient heritage from the ice itself, the earth shaper. Like the ice, we have been cruel to the face of the planet and the life upon it. A chill wind lingers about us. With a few slight exceptions we are merciless. We have invented giant, earth-scavenging machinery to do what the ice once did. Does this explain the nature of the man whose mind is lost among ancient pictures? No, not entirely. For again from that dim mental studio, he peers out upon modern pictures and transposes them as in some totemic ceremony.

I once visited a distinguished artist whose primary interest is landscape and buildings, not animals. He was giving me a little preview of his latest work. I glanced at a deserted farmyard containing a few abandoned wheels and a broken pump. "I see you have a fox's face hidden by the well curb," I said.

My friend jumped up and peered at his canvas. "No, no," he protested. "That is not a fox. I had no such intention."

"It's surprising," I said in turn. "I can still see it. Look," I tried to point. "See it now, the eyes, the ears? There, just over the wellhead, watching."

"It can't be there," cried the artist, starting to pace restlessly before his picture. "Damn it, man, that's just a rusty pump, boards, and old wire. I didn't put a fox there. I didn't."

"It's there just the same, Dan," I said. "It's the perspective.

You got him in somehow. Or he sneaked in. Stand over here and look."

Dan peered at me strangely. "I'm not going to look," he countered sullenly. "If I look, I'll begin to see him myself. I almost do now. I won't look. You're spoiling my picture. I won't show any more." He began to stack his pictures face to the wall.

"Dan," I said, "I'm sorry. I just thought I saw—maybe it's my eyes." I removed my glasses and waved them. "I thought you had actually intended—"

"Just forget it," he sulked. I knew he would never give me a private showing again. There are pictures that come to affect me like Rorschach tests, where the psychologists try to peer into your head by way of your interpretation of an ink blot. I begin to look and the blots turn out always to be animals. A picture of a tropical swamp painted by a friend now deceased once hung on my wall. The scene was admittedly weird, but finally I began to see so many little faces in the twisted stumps that the thing got on my nerves. I gave it away. Probably, however, all of the creatures merely retreated to my private storeroom to await the time they might emerge again.

This is the way of it, I think. One has just so many pictures in one's head which, after one has stared at them long enough, make a story or an essay. Beyond that one is helpless. Naturally, if one is sufficiently distracted, or has dreams, as I often do, one tries to jot something down as at least a brief reminder. This can be a very efficient device in the hands of a man like Thoreau. I practice it in these late years, but unless one records detail it can be folly. Leafing through the old notebooks of my busiest, most diversified years, I have recently come upon two mysterious notations as frustrating to me as the unexpected fox in my painter friend's farmyard. The lines run as follows and are obviously unrelated:

> Story of the Three Bloodhounds
> The power of the mice

Now the three bloodhounds, in spite of suggesting something out of Sherlock Holmes, must have been intended as a tag for some curious dog story I had heard. The quick note implied haste. Years had passed before I was able to read that line again. If the story lies lost in the room of memories, I have been unable to recover it. As for the power of the mice, whatever their power was, it has totally vanished. If I had my choice as to which notation I would rather be informed upon, it would be the latter. In each case I would venture that these lost phrases hinted at nothing so visual as a redheaded woodpecker or the eyes of a snow leopard. They are consciously literary and therefore they faded from my mind.

On the other hand, there is the case of Willy, whose life was so tangential to my own that he would never have drawn a line in a notebook. Nevertheless, Willy lives on because of a certain grandeur and pathos in his end. I never learned Willy's full name. He was a black who had the night shift in the garage of our apartment house.

I used to come home and see Willy leaning upon the little fence by the door of the garage. Across that fence was a pharmacy and a brightly lighted shopping center. I would speak to Willy and wonder what so fascinated him across the fence by which he lingered. Only after Willy's death did I begin to understand his final days.

Willy must have known he was dying, but, like most of us in humble circumstances, he went on tending to the dark subterranean garage until his final days. By carefully leaning upon the fence, he could see across into the domain of life. It was a small break for him in the brightly lit evening. The fence, and it was a rickety affair, had begun for Willy to mark the boundary between life and death. He knew he was on the wrong side of the barrier, but there was nothing he could do about it. In his final days he had come unconsciously to yearn toward the lights and movement across the way. There were telephone booths there in which boys called girls, or girls called boys. It was all exceedingly attractive to a dying man.

I doubt if anyone else remembers Willy now, but I do. I can look at the fence where the roses grow each spring and see him standing there, a worn black shadow. I am Willy's last recorder. Have I made sufficiently clear the burdens that a writer carries? I sometimes think that men and their thoughts are like jack-o'-lanterns upheld on poles at Halloween. They float and grin awhile before some dark unanswering window, and then, like hollow pumpkins, they are taken down, dismantled, and cast out. Poor old heads, there was only a small light in them and a rotund expansiveness that soon withered. Our own case is no better. But Willy still stands immovably by the fence. I can just make him out. It is a matter of seeing, like the fox in the painting. The rotten fence pickets have become a vast menacing landscape, but Willy refuses to depart. He exists in me, he watches.

CHAPTER 16

The Letter

I REMEMBER," he had written. That was in 1947, and my hand shook remembering with him. Over a quarter of a century has passed and I have not answered his letter. He was the closest of all my boyhood friends. And if, in the end, I did not respond to his letter, it was not that I intended it so. I sent, in fact, a card saying I was moving to Philadelphia and would write to him later. All this was true. In a little bundle of letters kept in a box and consisting mostly of communications from men long dead, I faithfully preserved his message. Only slowly did I come to realize that something within me was too deeply wounded ever to respond. In the first place, let me state with candor that if I have let nearly thirty years drift by—I last saw Jimmy Dawes when we graduated from grade school—I was forty-one when I received his own first recognition of my existence. He had seen some pieces of mine in a national magazine, *Harper's*, and remembered my name. Perhaps in a sense this balances the equation.

He wrote me as a successful officer in a huge corporate enterprise. He remembered a surprising amount about our childhood activities and he recalled something that touched my naturalist's memories and sent me groping to my book shelf. "I remember," he wrote, "all those squirmy things we collected in jars and buckets and took home to put in the aquariums you made."

He was right even if he was chuckling a little. But he was not content with these boyish memories. He persisted until he came to the explanation, unexpected by me, as to why, though we had lived on into young adulthood in the same community, I had never encountered him after our eighth-grade graduation.

I have said earlier that though in the western towns of those years poor children might attend school with those of another economic level, there came a time when the bridge was automatically withdrawn. This I accepted and never questioned, though much later when I passed my companion's home I used to look up at it a little wistfully. I had once been welcome there —I suppose, looking back, like a dog, a pet good for one's son at a certain stage of life, but not to be confused with the major business of growing up. No, I should not put it so harshly; his mother, his sisters were kind. I know that in their home I saw my first *Atlantic Monthly* in the traditional red-brick covers.

I rarely saw Jimmy Dawes' father, but I knew him as a contained, serious man of business. He ruled a healthy, well-directed household in which one knew that because of the wisdom of father everyone would marry well, be economically secure, and that each child was bound to live happily ever after. Actually, because of the pecularities of my edge-of-town status, these good people had probably stretched things a little to please their only son, since we played happily together. If I led him to adventures in the fields and ponds around the town, these were no more than an *Atlantic* essayist of the time would have thoroughly approved.

When Jimmy vanished from my ken I suppose I ached a little, but I was adjusted to the inevitable loneliness of my circumstances. There was a very large high school in Lincoln and I never saw him again. Looking back, it is possible he was sent elsewhere. I repeat that the separation was so neatly handled that I never thought of it as more than the usual process of

growing up, of being always the aloof observer, never partici-
pant, in the success of other families. No, if Jimmy Dawes had
written me after all those years merely to wish me well and remember our pond adventures, I would have answered. Not
effusively—one comes to accept one's place in life—but to con-
gratulate him upon his success, the fine children of whom he
wrote, and to thank him for his interest in my few ephemeral
essays. There the matter would have ended. I know he meant
well, but having grown up in that ordered household he proceeded, either in tactless condescension or, more likely, by way
of explaining a thirty-year silence, to examine his father's role
in the matter.

The social facts of life had left me merely grateful for a few
shared years and timid, occasional entrances into a home fan-
tastically different from my own. I had realized, even as a
youngster, that Jimmy Dawes was headed for something I would
never be. But now here was Jimmy Dawes, after thirty years,
telling me with no trace of regret that he had only pursued his
father's directive. Father had deemed it time that his son dis-
continue this enthusiasm for a pond-dipping fox-child and get
on with the business of where a properly directed young man
should go.

Papa had said it, apparently, right out loud. Oh, it is true that
at the end of the letter Jimmy Dawes had suggested, in a belated attempt to mollify my feelings, that perhaps he had himself been responsible in having taken up with all these matters
too enthusiastically. Doubtless father had been right. Fathers,
in his world, always were. He managed to tell me that also.

At the end of the letter I found myself shaken over what to
Jimmy was a simple fact that I would easily appreciate. Could
I now venture pretentiously to Jimmy Dawes, that I was, after
all, a social scientist? Could I say that I had long since accepted
all that he had to tell me, and then ask him upon what impulse
he had chosen to repeat what his father had said? And if you
chose to forget your friend, the nagging thought persisted, why

are you busy with this resurrection now? Am I made respectable at last by my printed name, because the *Atlantic Monthly* was taken and read in your father's house long ago?

I had liked Jimmy Dawes; that was why his belated emergence was so painful. There was still in that letter some trace of a bounding, youthful eagerness long lost by me. I sent the brief card of acknowledgment, went on my way to Penn, and proceeded to be haunted the rest of my life by his letter. No, not quite. In preparation for eventual retirement I began in the late summer of 1974 to destroy old files. I knew where the letter was, though I had not read it again in all those years. I read it once more, and because Jimmy had been my boyhood chum I reached for the telephone and asked for a number and an address far away. "Sir," the operator's voice came back, "no such name exists at that address."

At heart I knew it would be so, but at least I had finally nerved myself to try. His children would be grown and married. He would be retired and playing golf in Florida, or perhaps reading a vastly changed *Atlantic Monthly,* like his father, in an equally traditional, well-managed home. Slowly, deliberately, I tore up the long-cherished letter. It had served its purpose. Through it I began to remember where part of my interest in the living world began.

On the shelf where I had started to fumble when Jimmy Dawes' letter arrived was an old book, bound in green cloth with a stylized fish in gold stamped on the cover, a book bearing the unimpressive title *The Home Aquarium: How to Care for It.* A man equally obscure had published that book in 1902, five years before I was born. His name was Eugene Smith and whatever else he did in life I do not know. The introduction was written in Hoboken, New Jersey.

The copy I possess is not the one I borrowed and read from the Lincoln City Library while I was still in grade school. So profound had been its influence upon me, however, that in adulthood, after coming East, I had sought for it unsuccessfully in old book stores. One day, in my first year in graduate school, I had been turning over books on a sales' table largely strewn with trivia in Leary's famous old store in Philadelphia, a store now vanished. To my utter surprise there lay three copies with the stamp of the golden fish upon their covers. It was all I could do to restrain myself from purchasing all three. It was the first and last time I ever saw the work in a book store. The name B. W. Griffiths and the date 1903 was inscribed on the fly leaf. To this I added my own scrawled signature and the date 1933. Now, in the year 1975, I still possess it. I have spoken in the past of hidden teachers. This book was one such to me.

It is true there had been my early delvings in sandpiles, and so strong is childhood memory, that I can still recollect the precise circumstances under which I first discovered a trapdoor spider's nest. My amazing, unpredictable mother was the person who explained it to me. How, then, did this pedestrian work on the home aquarium happen to light up my whole inner existence?

There were, I think, two very precise reasons having nothing to do with literature as such that intrigued me about this old volume. Most of the aquarium books of today start with the assumption that you go to a pet store and buy tanks, thermometers, specialized aeration equipment, and even your assemblage of flora and fauna "ready made." Smith's book contained no such assumption. You got the glass, you cut it yourself, you made bottoms and sides of wood. Then, somewhere, you obtained tar to waterproof the wood and the joints. Moreover, Smith had given a simple running account, not alone of easily accessible fresh-water fish, but of local invertebrates with which aquariums could be stocked.

I genuinely believe that it was from the pages of his book that I first learned about the green fresh-water polyp *Hydra viridis*, so that later I identified it in one of my own aquariums, not from the wild. In other words, there had been placed within my hands the possibility of being the director, the overseer of living worlds of my own. If one has the temperament and takes this seriously one will feel forever afterward responsible for the life that cannot survive without one's constant attention. One learns unconsciously about ecological balance, what things may the most readily survive together, and, if one spends long hours observing, as I later came to do, one makes one's own discoveries and is not confined to textbooks.

This leads me to the other value gained from Smith's plain little volume. I spent no time, as a midlander, yearning after tropical marine fish or other exotic specimens. I would have to make my own aquariums and stock them as well. Furthermore, for a lad inclined as I was, one need not confine oneself to fish. One could also make smaller aquariums devoted to invertebrate pond life.

By chance I encountered Smith's book in midwinter and it would be a normal parental expectation that all of this interest in "slippery things" would have worn itself out while the natural world was asleep under pond ice. In the flaming heat of enthusiasm, however, I grew determined not to wait upon nature. I first secured some wood and glass scraps from a nearby building project and tar from a broken tar barrel. Though I had no great gifts as a carpenter I did the job by persistence from materials then strewn casually about every house under construction. I boiled and applied the warm tar myself. In one triumph, I even made a small aquarium from a cigar box.

In a few days I had enough worlds to start any number of creations. To do so I had to reverse the course of nature. Elders may quail but this is nothing for children. So it was winter? Snow lying thick over the countryside? Ponds under ice? Never mind. I knew where the streams and ponds were. I had also learned that many forms of life hibernate in the mud of ponds. All that was necessary was to improvise a net, again homemade, take a small lard bucket or two, and trudge off to the most accessible Walden.

The countryside was open in those days. On one visit to

Lincoln several years ago I thought it might be good to tramp out to that old pond where so many generations of boys had swum, waded, or collected. Forbidding fences warned me away. It was now part of a country club of the sort doubtless frequented by the successors of the parents of Jimmy Dawes. I speak no ill. If a country club had not acquired the ponds and landscaped the greens, all would have been filled in by suburban developments in any case. But this was all wild once, and the feeling that is left is somehow lost and bittersweet. The pond is there. It is not the same pond. It is "reserved." It has been tamed for rich men to play beside. Either that or the developers come. One takes one's choice. No. Not really. One has no choice.

On that winter day so long ago I almost lost my life. I arrived at the pond and chopped an experimental hole near the shore where I worked my clumsy mud-dredging apparatus back and forth. My plan was successful. I was drawing up a few sleeping water boatmen, whirligig beetles, and dragonfly larvae, along with other more microscopic animalcules. These I placed in my lard buckets and prepared to go home and begin the stocking of my little aquariums. A forced spring had come early to my captives.

There were skates on a strap hanging around my neck, and before leaving I thought I would take one quick run over the pond. It was a very cold day, the ice firm. I had no reason to anticipate disaster. I made two swift passages out over what I knew to be deeper water. On the second pass, as I stepped up speed, there was a sudden, instantaneous splintering of ice. The leg to which I had just applied skating pressure went hip deep into the water. I came down upon my face. I lay there a moment half stunned. No one was with me. What if the rest of the ice broke? Even if one held on to the edge one would freeze very quickly.

I waited anxiously, trying not to extend the ice-fracture by struggling. I was scared enough to yell, but it was useless. No one but a boy infused with the momentary idea of becoming a creator would be out on a day like this. Slowly I slid forward, arms spread, and withdrew my soaking leg from the hole. I must have struck an ice bubble with that one foot. The freezing weather fortunately permitted no general collapse of the ice. In one sweating moment I was safe, but I had to jog all the way home with my closed buckets.

After such an event there was no one's arms in which to fall at home. If one did, there would be only hysterical admonitions, and I would be lucky to be allowed out. Slowly my inner life was continuing to adjust to this fact. I had to rely on silence. It was like creeping away from death out of an ice hole an inch at a time. You did it alone.

Critics, good friends in academia, sometimes ask, as is so frequently the custom, what impelled me to become a writer, what I read, who influenced me. Again, if pressed, I feel as though I were still inching out of that smashed ice bubble. Any educated man is bound to live in the cultural stream of his time. If I say, however, that I have read Thoreau, then it has been Thoreau who has been my mentor; this in spite of the fact that I did not read Thoreau until well into my middle years. Or it is Melville, Poe, anyone but me. If I mention a living writer whom I know, he is my inspiration, my fount of knowledge.

Or it is the editor of my first book who must have taught me this arcane art, because if one writes one must indeed publish a first book and that requires an editor. Or if one remains perplexed and has no answer, then one is stupid and one's work is written by a ghost who is paid well for his silence. In one institution where I taught long ago, it was generally assumed that all of us young science instructors were too manly to engage in this dubious art. Our wives produced our papers.

I myself believe implicitly in what G. K. Chesterton wrote many years ago: "The man who makes a vow makes an appointment with himself at some distant time or place." I think this vow of which Chesterton speaks was made unconsciously by me three separate times in my childhood. These unconscious vows may not have determined the precise mode of whatever achievement may be accorded mine, or what crossroads I may have encountered on the way. I mean to imply simply that when a vow is made one will someday meet what it has made of oneself and, most likely, curse one's failure. In any event, one will meet one's self. Let me tell about the first of those vows. It was a vow to read, and surely the first step to writing is a vow to read, not to encounter an editor.

It so happened that when I was five years old my parents, in a rare moment of doting agreement, looked upon their solitary child and decided not to pack him off to kindergarten in that year. One can call them feckless, kind, or wise, according to one's notions of the result. Surprisingly, I can remember the gist of their conversation because I caught in its implications the feel of that looming weather which, in after years, we know as life.

"Let him be free another year," they said. I remember my astonishment at their agreement. "There'll be all his life to learn about the rest. Let him be free to play just one more time." They both smiled in sudden affection. The words come back from very far away. I rather think they are my mother's, though there is a soft inflection in them. For once, just once, there was total unanimity between my parents. A rare thing. And I pretended not to have heard that phrase "about the rest." Nevertheless when I went out to play in the sunshine I felt chilled.

I did not have to go to kindergarten to learn to read. I had already mastered the alphabet at some earlier point. I had little primers of my own, the see-John-run sort of thing or its equivalent in that year of 1912. Yes, in that fashion I could read. Sometime in the months that followed, my elder brother paid a brief visit home. He brought with him a full adult version of *Robinson Crusoe*. He proceeded to read it to me in spare moments. I lived for that story. I hung upon my brother's words. Then abruptly, as was always happening in the world above me in the lamplight, my brother had departed. We had reached only as far as the discovery of the footprint on the shore.

He left me the book, to be exact, but no reader. I never asked mother to read because her voice distressed me. Her inability to hear had made it harsh and jangling. My father read with great grace and beauty but he worked the long and dreadful hours of those years. There was only one thing evident to me. I had to get on with it, do it myself, otherwise I would never learn what happened to Crusoe.

I took Defoe's book and some little inadequate dictionary I found about the house, and proceeded to worry and chew my way like a puppy through the remaining pages. No doubt I lost the sense of a word here and there, but I mastered it. I had read it on my own. Papa bought me *Twenty Thousand Leagues Under the Sea* as a reward. I read that, too. I began to read everything I could lay my hands on.

Well, that was a kind of vow made to myself, was it not? Not just to handle ABC's, not to do the minimum for a school teacher, but to read books, read them for the joy of reading. When critics come to me again I shall say, "Put Daniel Defoe on the list, and myself, as well," because I kept the vow to read *Robinson Crusoe* and then to try to read all the books in the local library, or at least to examine them. I even learned to scan the papers for what a boy might hopefully understand.

That was 1912 and in the arctic winter of that year three prisoners blasted their way through the gates of the state penitentiary in our town. They left the warden and his deputy dead behind them. A blizzard howled across the landscape. This was long before the time of the fast getaway by car. The convicts were out somewhere shivering in the driving snow with the inevitable ruthless hunters drawing a narrowing circle for the kill.

That night papa tossed the paper on the table with a sigh. "They won't make it," he said and I could see by his eyes he was out there in the snow.

"But papa," I said, "the papers say they are bad men. They killed the warden."

"Yes, son," he said heavily. Then he paused, censoring his words carefully. "There are also bad prisons and bad wardens. You read your books now. Sit here by the lamp. Stay warm. Someday you will know more about people out in the cold. Try to think kindly, until then. These papers," he tapped the one he had brought in, "will not tell you everything. Someday when you are grown up you may remember this."

"Yes, papa," I said, and that was the second vow, though again I did not know it. The memory of that night stayed on, as did the darkness and the howling wind. Long after those fleeing men were dead I would re-enter that year to seek them out. I would dream once more about them. I would be— Never mind, I would be myself a fugitive. When once, just once, through sympathy, one enters the cold, one is always there. One eternally keeps an appointment with one's self, but I was much too young to know.

By the time of the aquarium episode I was several years older. That, too, was a vow, the sudden furious vow that induced me to create spring in midwinter. Record the homely writing of Eugene Smith, placing in my hand a tool and giving me command of tiny kingdoms. When I finally went away to graduate school I left them to the care of my grandmother Corey. I told her just how to manage them. She did so faithfully until her death. I think they brightened her final years— the little worlds we cared for. After the breakup of the house I searched for them. No one could tell me where they were. I would have greatly treasured them in the years remaining. I have never had an aquarium since, though expensive ones are now to be had.

I suppose, if I wrote till midnight and beyond, I could conjure up one last unstable vow—when at nineteen I watched, outwardly unmoved, the letter of my father crumble in the flames. I started, did I not, to explain why a man writes and how there is always supposed to be someone he had derived his inspiration from, following which the good scholar may seek out the predecessor of one's predecessor, until nothing original is left. I have said we all live in a moving stream, as surely as a catfish groping with its whiskers in the muddy dark. I have seized this opportunity nevertheless to ensure that my unhappy parents' part in this dubious creation of a writer is not forgotten, nor the role of my half-brother, who accidentally stimulated me into a gigantic reading effort. As for Eugene Smith, he gave me the gift of wanting to understand other lives, even if he almost stole my own upon that winter pond.

I would like to tell this dead man that I fondled his little handbook as I wrote this chapter. We are not important names, I would like to tell him. His is a very common one and all we are quickly vanishes. But still not quite. That is the wonder of words. They drift on and on beyond imagining. Did Eugene Smith of Hoboken think his book would have a lifelong impact on a boy in a small Nebraska town? I do not think so.

Ironically, one of the senior officers of the firm that published Smith's work asked me not long ago if he could interest me in a project and would I come to lunch. The letter was pleasantly flattering. I wrote the man that I would be glad to lunch with him, though in all honesty I was heavily committed elsewhere. After his original invitation, I was never accorded the dignity of a reply.

Sometimes this is called the world of publishing. It is a pity. I would have liked to tell this important man, over a cocktail, about a man named Smith whose book was published by his very own house before he, this generation's president, had been born. I would have been delighted to inform him that there was a stylized gold fish on the cover, and to what place the book had traveled, and how it had almost drowned, as well as uplifted, a small boy. Alas, this is a foolish dream. The presidents of great companies do not go to luncheons for such purposes. As for me, these strange chances in life intrigue me. I delight or shudder to hear of them.

I am sorry also, Jimmy Dawes, that your letter went unanswered. I genuinely hope that you and your grown-up family are happy. I apologize for the mind-block that descended upon me there in Ohio, that old psychic wound that should have been overlooked had I been stronger. But at the last I tried. Perhaps that is what my parents meant when they said, "There'll be all his life to learn about the rest." I learned part when my father died, the part about the cold. Jimmy Dawes was still in my home town, a few blocks from where I lived, but no note of condolence ever came from him or his family. I kept your letter, Jimmy. Almost thirty years later I tried to call you long distance. And now at last the wound is closing. It is very late.

CHAPTER 17

The Ghost World

IN the dark fall of 1948 my wife and I were living in a second-floor apartment in a private house in suburban Philadelphia. An outside circular stairway of iron descended to the backyard through which ran a little brook. Ideally we should have been happy but the great stone house was bleak and sunless, on the wrong side of a forest hill. We soon learned that our landlady, although a cultivated woman, had a habit of prowling the corridor and, in our absence, intruding into our apartment at will. Finally, I missed a book or two, no doubt not intentionally stolen, but appropriated out of aimless curiosity, and forgotten. I could hardly organize a search, though the loss was serious in terms of some studies I was pursuing.

We found ourselves huddling more and more into the kitchen and bedroom, for our living room adjoined the "listening" corridor. We would have moved into the city, but the city was still overcrowded from the war years. Landladies in the neighborhood of the university demanded bribes and, in any case, the area was deteriorating. So we hung on, enduring our petty troubles and hoping a better place would eventually turn up. The damp increased as autumn descended. I have always been unusually susceptible to the respiratory ailments that are so common in great cities.

One night, after what had seemed a mild cold, I awoke in the dark conscious that I was running a fever and babbling a lecture to some unseen audience. Slowly, as my consciousness steadied, I grew aware of something strange. Outside, lightning bolts sporadically split the dark. I could see through the bedroom window a torrential rain in progress. After each stroke of lightning I waited for the following thunder. There was none. I was deaf. The last lines were going down. I was alone with that knowledge in the dark.

"Mabel," I said, touching my sleeping wife. "Say something. For God's sake, say something." The lightning flickered again. I saw my wife start up. Her lips moved. I heard nothing, nothing at all. I arose and lit the lamp and stared at her.

Now to one not born to a stone-deaf mother this episode may seem trivial, at best a matter for a specialist who would soon set things right. We were, however, in a city about which I then knew very little, in spite of my graduate student days of over a decade before. On someone's recommendation I had gone, for treatment of my cold, to a nearby doctor of elderly status who had given me one shot of penicillin, then the all-purpose wonder drug. "This will fix you," he had said. Whatever the mixture, it almost did. I had broken out in great hives from head to foot, something penicillin has never done to me since. The doctor had not intimated that I should return and after one glance at my body I had decided he knew what he was talking about.

So now it was here, the thing so long feared, come in the night, creeping upon me in the midst of fever. I spoke to my wife, tugging futilely at my ears. "It's gone," I said desperately. "I can't hear you, I can't hear anything." We sat in the little kitchen, while she wrote me a note. "We'll find another doctor in the morning."

I nodded, but my confidence had been badly shaken. Besides, there was the economic situation to consider. I was a fully tenured professor but I had been at Penn for scarcely a year. What, after all, did they owe me? How could I teach without hearing? In all my professional experience I had known only one man who had managed it—a famous historian who had given most carefully organized lectures in a voice necessarily monotonous. We had passed written questions up to him if we so wished. His trouble had been hereditary and had fallen upon at least one of his children. I had been acquainted with him; he had been kind to me long ago. How, and by what help he had managed to obtain his post, I did not know. This was a harsher time, in a great industrial city. What could I expect of an institution where I had taught so briefly?

Dawn came as I paced restlessly about our little apartment. I tried to read, cast the book aside. As the rest of the world awoke and wires began to buzz, my wife consulted one of my colleagues. We obtained the name of a good general physician. We drove miles across to the Main Line, as Philadelphia still calls this important locale from the days when the Pennsylvania Railroad was a power, and wealth and family names and estates of eminence were concentrated there.

The doctor received me kindly, but his efforts were in vain. I could sense, though deafened, the uneasiness of his pacing around my chair. He struck a tuning fork alongside my head, tried other experiments that should have revealed any semblance of nerve response. All failed. I had come from the anatomy laboratories of a war. He did not need to tell me what he told my wife. "He needs a specialist," the doctor informed her.

"Please try," she said. He lifted the phone. Dr. Edwin Longaker, of Ardmore, could see me that afternoon.

We went home to wait. I wanted to be alone. I wandered down to the little brook where a faint November light played upon the ripples. I sat on a stump and threw a stone in the water. I thought of those derelicts, including myself, who had similarly waited in the Kansas City yards so long ago, the men cast off by the city, sitting with hands gripping and ungripping, or throwing cinders at a barrel. All those sweating years of danger and effort. If I had not come here this might never have happened. I might still be sitting in another hobo jungle. A tuning fork whose vibrations not even the bone would pick up had clarified this. My God, the nerve *must* be gone. "Face it, face it, face it. You're not going to hear, ever again." The tortured, straining features of my mother came back to me. How long before I turned paranoid, before—

My wife came to call me for lunch. I could see her anxious, labored effort to shout into my ear. Finally, she took my hand and led me to the stairway. "Christ," I thought, "is this what she is going to live with from now to the end of her life?"

"No," I determined, the pacing fox at the back of my mind beginning to hurl itself at an invisible fence. "No, she's not. Wait a little, wait and see. Remember what they told you in the Lincoln dispensary."

My classes for the day were canceled. The hours passed. We drove to the second doctor. I remember still the quick, efficient courtesy. This time my eardrums were carefully examined in the head reflector. A fork was struck expertly.

I nodded. "Yes, yes." This time I had picked up the vibration. The auditory nerve was still functioning. Again. Again, I nodded. Sweat was dripping from my forehead. Shut out hope, shut out fear. It's too early for either. Wait, just wait. In that little dispensary they had tried to kill you with fear; this time it may be by hope.

I was handed a glass of water and made to understand I was to gulp it at a signal. My nose was grasped and the tip of a tube inserted. At the signal I swallowed. In the same instant a perfectly coordinated hand had driven a blast of air against my Eustachian tubes. Something was momentarily torn open, air pressures equalized. For the space of a second or two I could hear my wife and the doctor exchange perfectly intelligible remarks. Then the veil descended. They both looked at me.

"I don't" I said, trying to be measured, " I don't want to

sound over-optimistic, but for just a second there I could hear you plainly." I repeated the words. I felt like a ghost trying to announce his presence in a convincing manner. I saw my wife throw her arms around the doctor. I saw him push up the reflector above his eye with quiet confidence. I looked on again in doubt from behind the veil.

"Otitis media." They scrawled it out for me. "Heavy infection of both middle ears. Eustachian tubes closed. Unequal pressure against the drum. It may take as long as six months, but you'll get your hearing back. Back, you understand? But it will take time and many visits."

I got up. I thanked the doctor blindly. He was a very skillful man. I was not to know then that he would be my nose-throat specialist until his death in 1973, still practicing in his eighties, one of the last of the old breed of physicians. We came to know each other so well, and he pursued his profession so avidly, that if either my wife or I needed attention he would take us on Saturdays and Sundays. Doctor Ed, we called him by then. I will remember him all my life.

But to return. It was a weary, weary time. I had to get others to take my classes while I minded the office and handled correspondence as well as I could. My then secretary, Madge Lambertson, whose husband had fallen at Okinawa, protected me from intrusion as much as possible.

Twice a week I journeyed the miles to Dr. Ed's office to have the swollen Eustachian tubes reopened and the drum massaged. For a few lingering minutes the treatment would hold and I could hear, after a fashion. Then the curtain would descend once more. I would come in to Thirtieth Street Station in Philadelphia on the interurban train. All around me chatting, laughing people on their way to work passed me like ghosts, only their gesticulating hands and moving lips proclaiming the life within them. The sounds of trains, of rolling baggage trucks, did not exist for me. A vast silence reigned in that busy place. I slid like an observant ghost from pillar to pillar. I did not wish to be noticed by colleagues lest I be engaged in conversation and have to confess my utter inability to hear.

One of my older married students discovered my problem and tried to help. "My husband," she explained, "is a research chemist. He is working on a new drug that is effective against viruses that penicillin will not touch." She was correct, as it turned out, and the drug was later to prove very successful, though I no longer recall which of the many "mycins" it was. There is sometimes a long step between the laboratory and the medical profession at large. The drug was beginning to be mentioned hopefully in the research journals, but it had not yet been placed on the market. No physician was authorized to use it. Looking back, I suppose by signing the proper waivers I could have become a guinea pig, but no one suggested it, nor, I am sure, would Dr. Ed have advised it. Nevertheless, it and its accompanying variants were destined to come into favor a few years later. In the meantime my well-meaning student was frustrated.

While I floundered in utter silence, a scientifically oriented magazine which had requested an article from me upon human evolution reneged in favor of a more distinguished visitor to America. Whether or not the editor realized it, I had counted heavily upon that piece. I had researched the subject considerably, but now I turned aside from the straitly defined scientific article. I had long realized an attachment for the personal essay, but the personal essay was out of fashion except perhaps for humor.

Sitting alone at the little kitchen table I tried to put into perspective the fears that still welled up frantically from my long ordeal. I had done a lot of work on this article, but since my market was gone, why not attempt a more literary venture? Why not turn it—here I was thinking consciously at last about something I had done unconsciously before—into what I now term the concealed essay, in which personal anecdote was allowed gently to bring under observation thoughts of a more purely scientific nature?

THE GHOST WORLD

That the self and its minute adventures may be interesting every essayist from Montaigne to Emerson has intimated, but only if one is utterly, nakedly honest and does not pontificate. In a silence upon which nothing could impinge, I shifted away from the article as originally intended. A personal anecdote introduced it, personal material lay scattered through it, personal philosophy concluded it, and yet I had done no harm to the scientific data. I sent the piece to one of the quality magazines, which accepted it. Out of the ghost world of my journeys through the silent station arose by degrees the prose world with which, it is true, I first toyed long ago, but which had been largely submerged by departmental discipline. *The Immense Journey*, perhaps my most widely read and translated book, was born on that little kitchen table where my wife had to write me notes to save her voice.

I had lived so long in a winter silence that from then on I would do and think as I chose. I was fond of my great sprawling subject, but I had learned not to love anything official too fondly, even high office. One had to stand aloof. Otherwise one was easily destroyed. This is also true of the writer beset by his own temptations.

Perhaps I had begun to learn independence among the mad Shepards, or freezing in midnight streets, or listening to my father declaim "Spartacus to the Gladiators," or when he coiled his fist and made me shiver when he read from Shakespeare:

> "He was a kinde of Nothinge
> Until he forged himself a name."

The months of winter passed, along with bitter disappointment. I wrote in silence, dreaming of digging days under the badlands sun. I wrote only to entertain myself, to keep the shadow back of me. Finally, while waiting for supper one late March evening, a soft sputtering purr seemed to emanate from somewhere near the stove.

"Mabel," I shouted, leaping up, "the gas flame, is it going? The gas flame, I heard it! I heard it!"

I buried my face in my hands. Yes, there it was, the little sound under the kettle. I looked up. "I will never be angry at noise again, not ever in my life," I vowed brokenly. It was a vow no one can keep for long in an industrial civilization, but I was returning from the dead world that passed me every morning in the railroad station.

Understandably, all was not yet well. There were hours when my wife and I could converse reasonably. There were days, damp days, when the mist still closed in about me. There were days when I discovered by chance that I could open the tubes and balance the air pressure at the ear drum only by lowering my head over the edge of the bed and listening upside down. A fine spectacle for class, I thought grimly. Professor Eiseley receives a question lying with his head hanging over the edge of his desk. Great for the campus photographers. I should have been a guru of the sixties. Everything would have been acceptable then. The students might have taken it up, head-dropping.

"It's nothing," remarked Dr. Ed. "A drop of moisture rolling around in there."

"How do I get it out?" I asked nervously. "I don't like to hear upside down."

"You'll have to wait," he counseled. "Give it time. Spring is almost here." He smiled kindly. "You've had a long siege."

"I can use the phone now," I said proudly. "Nobody guesses that I have any hearing difficulty."

A few nights later I stepped out onto our little balcony. The air was soft. Clear as the voices of my childhood, the spring peepers sounded in the trees above the brook. I listened as though I could never have enough. "I ought to write something about this and the kettle," I said to my wife. "But mostly the kettle. I mean that little flame and how it purred. People don't appreciate things like that, they never do till they are gone. I turn it on now sometimes when you're away, just for company."

"You'd better keep that to yourself," she said smiling, "or tell Doctor Ed. Nobody is going to understand you."

(179)

THE GHOST WORLD

"Someday I'll tell," I said, "about the ghost world at the station. The people were like shadows. They walked and they never made a sound. I'll never see that again. At least," I hesitated, "I hope not."

Twenty-six years have gone since that spring when the gas flame muttered and the peepers sang in the night. But there have been times of trouble. If I have a bad head cold it is liable to strike straight for my Eustachian tubes. It is perhaps folly for me to live in this Atlantic climate, but I have grown used to certain amenities and the deserts are not the deserts of my youth. There are lakes in some of them, retirement communities, speed boats, and water skiers.

I mope restlessly about and consider the matter. Dr. Ed is gone so recently that I have not had the heart to find another doctor. I like men who die in harness. The last time he attended me, he was standing up through sheer will power. He tripped. I had to hold him.

"Take care now, Ed," I told him, gazing at a wonderful old photograph taken down a long lane with a rail fence and a house in the distance.

He'll be walking that road soon, I thought, to the big house of his childhood. I raised a cheery hand, sweeping my eye for the last time over the little sterilizer for instruments, the stools, the reflector on his forehead, probably the same one with which he had first examined my ears. "Ed," I began, and thought better of it. Neither of us knew how to handle such things. "Don't ever retire," I said. "We need you."

"Not likely," he grunted. In a week he was gone.

Personally I have no compass, no directions. For me there is no clear stretch of road, but for him the picture will be hanging there. He will be all right. He will know where to go. I suppose that is why he kept it on his wall. Most of us lose our way when the time comes.

CHAPTER 18

The Dancers in the Ring

H E was a man younger than I and he came to me across the party floor with the sure arrogance of one determined to stab me verbally because he wasn't the kind to suggest we go outside. Moreover, this would bring him pleasure. My friends would hear.

"Eiseley," he said loudly, with seeming solicitude, "you ought to get a grant. You ought to go out into the sun. Learn what it's like in the heat and dust. It would help your writing. The big spaces would do you good. You need practical experience."

I looked at him. What did it matter now? I knew what animus had always consumed him, for reasons unknown to me. We were not intimates. He could afford this. I couldn't. I could feel a slight tremor in the glass I turned very carefully in my hand. I was one of the hosts at this meeting. As he spoke, I thought of the respirators we had worn under the sullen mountain, where the dust could provoke silicosis.

"The dust?" I said.

"Yes," he jibed.

"It helps you write?" I queried. He eyed me with a ferocious uncertainty. I turned the glass again. I fixed my mind on that little rat-haunted cabin in the Mohave and kept it there.

(181)

"The heat, too," I meditated. "Yes, the heat. Certainly I must apply for a grant. Kind of you to think of it, before I retire."

In reality, there is a dust one may breathe among old books which can be just as fascinating, the heat as infernally oppressive, as any amount of crawling about in tombs and deserts. Let me explain how my own adventures in those obscure regions began. Sometime in the very early 1950s I was approached by a New York editor who wanted me to write a book about the Darwinian epoch. I remember that I was taken to lunch in the customary way and that we were seated, by chance, in a quite noisy place during a time when I was still recovering from my experience in the ghost world of deafness.

The young editor was soft-spoken, and from the fragments of our conversation I pieced together the notion that an unusually well researched book was wanted. The subject interested me. I had taught classes in evolutionary history. Overwhelmed with the confidence expressed in my abilities, I went home and began the collection of data—something that in the case of such a monumental project demands time and the utilization of original sources. Moreover, duplicating machines were not then common in libraries. One slowly copied one's material by hand under quite difficult circumstances.

At that time the huge air-conditioned Van Pelt Library at the University of Pennsylvania was not in existence. The old Furness Library was overloaded, lacked air conditioning, and was grimed, particularly in its less-used stacks, by the soot and dust of a century. The lighting in some sections was so poor that the use of a miner's lamp would have been justified. Desiccated red leather bindings had a way, in summer, of ineradicably staining one's suits or collapsing into mummy powder that one breathed. No great effort had been made to treat or renew such bindings for decades. As a result a treasure trove of inestimable proportions was slowly crumbling away. It was not until some years later that the administration, of which I was

then a part, began the long-belated and expensive effort to restore the books.

Contrary to the expressed opinions of the man who ex-
pounded upon the benefits of heat and dust in the field, I
think I have never endured more unpleasant conditions than
in those ancient library stacks. I have dug fossils in places
where the sun glare demanded dark glasses, but the air was
clean. I have swung long-handled spades from trenches fifteen
feet deep, where cowboys and cattle both came to stare be-
musedly as dirt puffed upward from below ground. All this
was in dry open country; it was not done in the dank humidity
of a Philadelphia summer at a time when the faculty was confined by custom to much more formal garb than is now the
case. No, my experiences in those stacks were closer to the subterranean experiences of the Egyptian tomb robbers.

But the treasures. Let us come to the point. The treasures
are in the mind that seeks them. Otherwise they are not recognized. Foreknowledge and preparation are needed before one
blinds oneself in dark passages or wearily runs a dusty thumb
down the smallest news notes in some ancient and crumbling
periodical. Penn is one of the oldest universities in the United
States. Some of the materials accumulated in its grimy library
were priceless but uncatalogued; sometimes a precious pam-
phlet would slip and fall through several floors of stacks, never
to be found again; sometimes under the grime were the features of a real Rameses the third. James Hutton's *Theory of the
Earth,* for example, I found to be lying there. It is now in the
rare book room. Already in that time it would have brought
one thousand pounds or more at Sotheby's in London. There
were other remarkable things.

The scholar who descends into the catacombs of the past is
endangered; he may lose his way. A given period, or a millennium, may become more real than the century he inhabits in
the flesh. The documents may prove extensive; thread leads
on to thread, passage to passage. My card files thickened with

material tangential to the purpose with which I started. Similarly the dedicated archaeologist, laboring under an enormous compulsion, may hasten to the next tumulus carrying his discoveries only in his mortal head, while disease or a viper under a stone can suddenly erase his achievement. Equivalent perils confront the delver into libraries.

Several years of toil passed with every spare moment occupied in a way that would have astounded and troubled the brisk young editor. I was becoming insubstantial, the present with its publishing schedules a bothersome encumbrance. My laboriously scrawled cards, which only I could decipher, continued their extension through fireproof cases that still line my walls. At this point two things threaten the researcher.

First, he may become so lost below ground, trail leading on to trail, that he may never emerge to publish. He may be stricken by a phobia of incompleteness. He may become a perfectionist who will not set pen to paper until he has consulted every document of the century in which he has come to live. The past has infiltrated his arteries and his brain; he no longer has a sense of mortality. He has lost the realization that the flesh-and-blood inhabitants of his chosen period went about their affairs as living and limited human beings.

No man could possibly assimilate every lie, half truth, and truth that bewitched the minds of a past century. With the relative clarity of aftervision we can attempt, at best, only some insights, some relative comprehension of ideas which will always be appraised anew by later generations. So great is the lure of documents, however, that it is easy to be lulled into a false sense of omnipotence. The drone of that buzzing fly, the publisher, recedes into the distances of the future we have unconsciously abandoned. The dust of the catacombs gathers upon our skins.

Second, as the accompaniment of this retreat, we may no longer care to organize this precious knowledge or fix it into a pattern. As in the case of the archaeologist protracting beyond reason his labors upon the city that the jungle has engulfed, publishing comes to seem a heresy. The final roadway sunk in the bitter lake has not been traced, a dead king may still lie hidden beneath the next unexcavated mound. Publish? There is not time. We will guard the secret in our heads.

What if it dies there? Send the foundation, or the museum director, a preliminary report just sufficient to sustain interest. The rest is sacrilege. Let us rest with our laborers in the shade. We are slowly abandoning the civilization that we knew. Here, for example, are the words of that great Egyptologist Flinders Petrie. "I here *live* and do not scramble to fit myself to the requirements of others. In a narrow tomb with the figure of Nefermat standing on each side of me—as he has stood through all that we know of human history—I have just room for my bed. Behind me is that Great Peace, the Desert." Dreaming in the dark corridors of the old Furness Library, I too had about reached Petrie's conclusion. I emerged ever more slowly into the light. The publishers had now waited several years. I am sure that they had given up all hope of receiving a manuscript.

As it turned out, they were wrong. The centennial celebration of the publication of *The Origin of Species* was approaching. I owed the publishers for their advance. Unfortunately, since the library did not remain open all night and I had classes to teach, I could not escape like others into the desert or the rain forest. In the end I drew upon my enormous files and wrote *Darwin's Century* to discharge my obligation. I wrote till my eyes wept from stress. I slept when tired without reference to day or night, and arose and wrote again. Since my wife was visiting her parents, I had my meals from the icebox and scuttled about the halls of my apartment house like the half-wild creature I had become. A few months after publication, the book received the national Phi Beta Kappa Award in science, and in the meantime I had been hailed forth from my subterranean haunts to assume the provostship.

The history of science is as full of abandoned sinkholes as a cavern. Theories emerge, have their moment, and vanish or, on the other hand, are slowly transformed into greater syntheses. Sometimes ideas regarded as impossible have a way of conquering a later generation. The theory of the drifting continents, once dismissed in professional geological circles, has emerged triumphant today.

The layman's idea that we pursue a sustained, undeviating march toward some final truth has slowly given way to the realization that, as in the case of all institutions, the history of science is beset by ambiguities, fears, and trends which may play upon and influence severely disciplined minds. Ideas do not spring full-blown from a single brain. There has to be wandering along bypaths, midnight reading, and sustained effort. Even chance may play a role, as in the original discovery of X-rays.

Or, as with the rise of the industrial city, more and better beef had to be driven from the country into places like London, "the great mouth." Quality wool had to be introduced in quantity for clothing. Far-seeing farm breeders undertook to supply improved strains of cattle and sheep long before the rise of scientific genetics. Thus in the eighteenth century men glimpsed the power of artificial selection to transform domesticated animals. One breeder speaks of such selection as a "magician's wand" summoning into life whatever form the breeder chooses. Many of these animal-inventors would doubtless have been shocked at Darwinian ideas of unlimited transformation throughout the geological past. They were, nevertheless, revealing a latent dynamism in living organisms, a capacity for physical change which would eventually constitute a newly structured biology.

As one gazes in retrospect upon the swirling cloud wraiths which closed the eighteenth century and ushered in the nineteenth, it becomes apparent how difficult it is to assign so intricate a scientific alteration of our world view to a single man magnified beyond human proportions. More recently the intellectual historian has come to see that scientific models of reality are remarkably complex in their origins and in their manner of transformation. While in no sense denigrating the achievements of the great masters of science, we have to recognize the truth spoken by the first of the atom breakers, Lord Rutherford. "It is not in the nature of things," said Rutherford, "for any one man to make a sudden violent discovery; science goes step by step and every man depends on the work of his predecessors. When you hear of a sudden unexpected discovery —a bolt from the blue, as it were—you can always be sure that it has grown up by the influence of one man on another, and it is the mutual influence which makes the enormous possibility of scientific advance." These are the words of a scientist, but very similar and equally modest expressions have emanated from the great artist Joshua Reynolds. Jacques Loeb, the brilliant experimental physiologist, was fond of quoting the dictum of his equally distinguished botanical mentor, Julius von Sachs, that "all originality comes from reading."

Sir Francis Bacon once spoke of those drawn into some powerful circle of thought as "dancing in little rings like persons bewitched." Our scientific models do simulate a kind of fairy ring or magic circle which, once it has encompassed us, is hard to view objectively. Truth is elusive. Perhaps William James put things most felicitously when he said, "The greatest enemy of any one of our truths may be the rest of our truths."

Within the magic ring, in other words, may be a truth we come to accept as the whole truth, just as physics became a closed system based upon a substantial, unalterable particle, the atom. Toward the end of the nineteenth century, certain workers within that particular fairy circle discovered radioactive elements that pointed toward atomic decay and disintegration. Perhaps, indeed, there was no atom in the old sense at all. Instead, the supposed truth began to erode in the face of new evidence until finally all the younger dancers had moved across into the circle of the new physics now current among us. To use James' analogy, the discovery of other truths destroyed the reality of the one truth under whose influence the bewitched researchers had long been dancing.

By now I may well be queried as to where all this talk of fairy rings and bewitched dancers is leading us. Let me answer in this way: I am describing figuratively what I learned in the grimy cellars of the old Furness Library, where I had long stumbled back and forth between the fairy ring of fixity and that of organic novelty, as those two circles were beginning to interpenetrate each other in Charles Darwin's young manhood.

The dancers bewitched by stable form came under the shadow of a new truth we now call evolution. Eventually they became the magical transformers evoked long ago by the back-country breeders. One can always bravely defend one truth. When contradictory truths multiply, one is forced to recognize a certain mockery written into the very fabric of nature. Our eyes would have to possess as many facets as those of an insect to perceive at once the relativity of truth itself. Bemused by that old century behind us, I had followed the dancers from one toadstool ring to another. I observed the transition and as I did so I became ever more conscious of the forgotten men who work to produce change before change comes about.

Sometimes the man achieving a great synthesis has the misfortune to be premature in his appearance. Then, as is exemplified by Gregor Mendel and by the founder of the hypothesis of continental drift, Alfred Wegener, he will die unnoticed in the cold and be lucky if a later generation recognizes him. He may not live to realize his own importance; he may, in fact, pass away ignored, ostracized, or persecuted. In the case of Wegener, his death on the Greenland ice cap may in the end have been suicidal. His problem, like that of Mendel, was augmented by the fact that he was not a "professional" in the field of his innovation. Charles Darwin remained always the wealthy, independent, inspired "amateur," happily before professionalism became fashionable. Professional academic science tends to strengthen the mutual pull of the dancers already circling in the ring, not, on the whole, those trying to dance out. As we become conscious of this fact, we may be able to combat instilled scientific conservatism more successfully.

Darwin came at a propitious time in a century of great change. Yet even he constantly put off the day of publication. In spite of the diversionary emphasis later placed upon his Odyssean voyage, he was in reality a great searcher of other people's works, another subterranean, if giant mole.

Darwin's recognition of others coming close to the significance of natural selection was always meager. For example, he wrote to the eminent geologist Charles Lyell in 1846: "With respect to variation I have found nothing but minute details scattered over scores of volumes." And, in reference to natural selection, he wrote to the natural philosopher Baden Powell in 1860 that "[I] received no assistance from my predecessors," a doubtful statement coming from a man with the appetite of a cormorant. Not for nothing did Reynolds Green, the botanical historian, observe in 1914 that Darwin "had a wonderful capacity of mastering the work of others and building upon it."

Like the man hidden in the rain forest, Darwin's secret had become an obsessive burden. He labored alone; he even left a will trying to divert his task to others. He suffered from doubts and hypochondria. Still he went on with his solitary excavations. He was never to possess the direct frankness of Lord Rutherford or Loeb. He affected not to care for priority and yet it is evident he cared very much.

I do not intend this chapter to be a detailed account of such difficult matters, but only to reveal the passions of the excavator, including myself. I had long been aware of this curious ambivalence in the mind of the great protagonist of natural selection. After writing *Darwin's Century*, I had suspected that the man who was perfectly informed upon the work of those who wielded the magician's wand over cattle must somewhere have received a hint, a glimmer of truth from a source beyond himself. This is not a denigration of great genius. Lord Rutherford merely stated a common but reluctantly acknowledged truth. After all, had not Darwin on one occasion, and admittedly in another context, given voice to the inscrutable comment: "to him who convinces belongs all the credit"?

This was an unconscious revelation from the very soul of the excavator. He pondered the works he unearthed but he could be uncommonly ruthless. Like a cannibal, he sometimes tore volumes apart to get at their interiors. The man in the buried ruins was not a simple innocent; he had a far-reaching, contemplative mind, but it was a mind of many moods and facets. In the immense correspondence of the great scientist with men of differing beliefs and attitudes, one can observe both the gracious ability to temporize, to remain opaque, or to retreat to the secrecy of the Victorian couch in a countryside unsullied by the intrusive telephone of the twentieth century. The man chose seclusion; it was his right. Many things perished in that solitude.

Shortly after the publication of *Darwin's Century*, my contemporary Bentley Glass, then at Johns Hopkins, asked me to contribute to a volume he was editing upon Darwin's forerunners. Looking back, I should have confined myself to a man I had already discovered and mentioned in my own book— namely, the American J. Stanley Grimes.

Passing hastily along a little-touched shelf in the murky depths of the library, I had seen what appeared to be a gleaming new leather binding among books of an older vintage. All were unclassified recent gifts. As my finger poised over the supposedly newly backed book, I realized that what I had thought was shining leather was actually the back of a giant cockroach perched upon and nibbling an ancient binding. As the insect scrambled away, I picked up the book upon which I could dimly make out the word "geology." There it was, *Phreno-Geology: the Progressive Creation of Man, Indicated by Natural History and Confirmed by Discoveries which Connect the Or-*

ganization and Functions of the Brain With the Successive Geological Periods, Boston, 1851.

I took it home, fascinated to discover that the book was rare. Even though the central theme of Grimes' work is somewhat entangled with the then-current enthusiasm for phrenology, it is a genuine precursor of certain ideas upon brain evolution and natural selection. Apparently intimidated by public feeling, Grimes lost his nerve; he recalled and destroyed much of this edition. Through some accident, a surviving copy had been presented to the Furness Library, its rarity having gone unnoticed. I read the book with avidity. Besides his speculations upon the brain, Grimes proved to be another dancer at the very edge of the old ring of fixity, another less sophisticated thinker already toying with natural selection. What today we would call mutations, he termed "idiosyncrasies," but the selective element is clearly there, since he speaks of his mutative elements as being "able to sustain the shock of new circumstances and survive."

I recorded but did not particularly elaborate upon Grimes. The reason was simple. His book predates the *Origin* by some eight years. We know that according to Darwin's own words he formulated his views in 1838. Grimes was one of the dancers who was prepared to dance out of the ring of fixity. To that degree he is symptomatic of the changing intellectual climate of the mid-century. He does not possess the fascination of Darwin's contemporary, Edward Blyth, because he cannot be directly linked with Darwin. Edward Blyth can be so linked.

I stumbled figuratively over the sarcophagus of Blyth in my renewed delvings into aging files of periodicals. I believe I may say that I resurrected him sufficiently that considerable energies at Cambridge and elsewhere have been devoted to laying his ghost, not with entirely satisfactory results. I came upon him, ironically, after *Darwin's Century* was published, and too late, because of the intricacies the trail presented, to elaborate my conclusions among the papers edited by Bentley Glass. This I regret, for it would have made my observations more generally available. I do not here intend to detail them because they are to be found in the *Proceedings of the American Philosophical Society* for 1959. It is merely the excavator's adventures which need concern us.

Basically I had thought I was through with Charles Darwin. I was turning to the collection of data for a second volume of evolutionary history, the part leading forward from 1900. As I worked patiently through the periodical *Nature* for 1911, I chanced upon the letter of an H. M. Vickers drawing attention to an 1835 paper of young Edward Blyth, in which the latter dwelt at some length upon artificial selection and then applied the principle to wild nature, just as Darwin was later to do. This would have drawn no deep interest on my part except for three things: First, the date and a following paper of 1837, fell within the time when we know that Darwin, convinced of the reality of evolution, was searching blindly for the means by which it came about. Second, the magazine in which these papers appeared was one taken and perused by Darwin even on the voyage of the *Beagle*. The work of his own friends appeared occasionally in it. Third, it struck me that Darwin's eldest son, who had been mildly questioned by Vickers, had laid something of a red herring across the trail by expressing to Vickers some question as to whether this was the man his father knew.

Knew? I almost soared out of my chair. Edward Blyth, the only well-known English naturalist of that name. A man who was footnoted and praised for a myriad factual details in Darwin's *Variation of Animals and Plants Under Domestication,* almost as though the latter were striving indirectly to make amends for some deed done long ago. A man who had corresponded with Darwin. I peered out from my catacomb where the dusty bulbs still flickered. I knew that my curiosity was going to bring down on me the wrath of those who regarded Darwin as a sacred fetish, a scientific saint who had appeared with the tablets directly after a mysterious world voyage, an Odyssean adventure. Darwin was the white knight of the evolutionary cause. I had long maintained that a world voyage, however romantic, was not necessary to demonstrate the reality of evolution by natural selection. The pieces of the puzzle were already lying about in libraries. What was needed was a shift in theological orientation and a great excavator. Darwin proved to be the latter.

That a preliminary oceanic solitude may have been beneficial I will not attempt to argue, but Darwin has placed himself on record with Huxley: "I have picked up most by reading . . . numberless special treatises and *all* agricultural and horticultural journals; . . . a work of long years"—the mole again surfacing after almost a quarter of a century, in the year of the *Origin.* Compare this remark with those directed to Lyell and Powell, "minute details," "no assistance from my predecessors." Once more, there were observable the evasions, the paradoxes of a basically good man who wanted, and didn't want, to publish, who desired fame and hid from it, who at one time found the Grail in the Galápagos and in another year among the crumbling volumes of great libraries. One must have sympathy; his problem lay partly in the political and theological nature of his time and the fears induced by it.

From his more cautious paper of 1835, Edward Blyth's thinking progressed to the point where in 1837 he concluded, meditating upon adaptation and selection: "to what extent may not the same take place in wild nature? *May not then, a large proportion of what are considered species have descended from a common parentage?*" This was the pattern of thought Darwin was to pursue through the rest of his days. Nevertheless we must remember that as late as 1836, after his return from the famous voyage, Darwin was still baffled. "If one species does change into another," he argued in his early notebooks, "it must be *per saltum* [a direct leap] or species may perish." Blyth expressed the view that such giant leaps would perish. No one has explained what changed young Darwin's mind, but in the very period of Blyth's publications his own view shifted to the opinions held by Blyth and was never afterwards altered.

The gist of my conclusion came to this: Darwin opened, shortly after his return home, notebooks on the species question. He claimed somewhat cryptically to be making progress. In the year of 1837 Edward Blyth produced a second paper elaborating his suggestions upon selection in wild nature. Then he asks the question already cited and proceeds in the end to doubt its possibility.

When I first read Blyth I took his disclaimer seriously, though it would not have affected the stimulus he had provided the great biologist. In later years, after more extended reading, I have come to realize that such disclaimers in that time were frequently a ritualistic gesture. French thinkers like Lamarck were in disrepute in England, which had grown markedly conservative after the French Revolution. Lamarck was an evolutionist. Therefore, say what you have to say and then disavow evolution. It was a common device. One must remember that even the wealthy Darwin procrastinated through some twenty-two years of uneasy but happy burrowing until the temper of the century had changed.

In short, these meaningless disavowals were often a bow to conformity in a society still theologically oriented and about as suspicious of French atheism as our own society became of "reds" during the McCarthy era. One cannot dismiss Blyth's influence upon Darwin because a poor young man in grave financial straits, and seeking a curatorial post, did not go out of his way to be emphatic. No, the coincidence is simply too detailed to explain away—the journals in which the searching Darwin could not avoid seeing Blyth's papers. But Darwin used Blyth for specific purposes while relegating to obscurity the vital keys that had changed the world of the existent to the world of potential organic novelty.

Faintly the words of young Blyth whisper in our ears after

the passage of one hundred and forty years. "I . . . hope," he wrote wistfully, "that this endeavor will induce some naturalists, more competent than myself, to follow out this intricate and complicated subject into all its details." A hint perhaps? Be at peace, Edward. The man you sought came, the man who said, "to him who convinces . . ." And I, who unearthed your whisper from the crumblings of the past, have been here and there excoriated by men who are willing to pursue evolutionary changes in solitary molar teeth but never the evolution of ideas.

We still dance in the rings, Edward. Let that be understood between us. You, too, were a solitary. Let the leaves come down once more. I have sat in many gatherings, moved in the world of scholars, read by lamplight those who preceded me, sought patiently for truth. My favorites, my midnight companions, have been consulted. DeQuincey, the little five-foot elf, did not find the ultimate secret in his night wanderings, or if he did was silent; Coleridge not in opium nor in the spilled sheets and books upon the floor; nor Sir Thomas Browne in burial urns; nor I in science. We round back, we return. Suppose we were seated, like the tramps we are, at a fire by the railroad. What would we say as the dark closed in—Men beat men, verbally or physically—is that the most of it?

No, not quite, Edward Blyth. For you see, I knew what would happen if I did not close the book upon you there in the catacomb. I reopened it for your sake, left it open for the world to read. And what happened afterward does not matter at all, any more than your ritual disclaimer. We are dancers in the ring.

CHAPTER 19

The Spectral War

I HAVE written of the fact that long ago in childhood I had cast dice on the floor of a ruined house and at evening fled away down a leaf-sodden road from a farm in which had once lived someone with my own name. What became of that family? Why was the house in ruins? Short though the white man's history may be in these western towns, it is sometimes terrifying by its very evanescence, fallen gravestones lost in wood-lots, houses sagging into empty cellars. In my later years I have always walked past gambling houses. I know by instinct that I am playing at harder and more dangerous games every day in the week. And speaking of madness in a kind of belated way, I, who have experienced the winds of Montana and watched sandstorms race over the deserts of Sonora, have a great respect for another unseen player, the wind.

The medical observers of the early century had a good deal to say about the life of women in soddies, on lonely homesteads, and what it could do to them. Americans made a mistake they have been paying for ever since. In response to the Homestead Act they have been strung out at nighttime into a vast solitude rather than linked to the old-world village with its adjoining plots.

I have lived under such circumstances. I have seen beaten, toil-worn women staring into immensity, listening to the wind. We were mad to settle the west in that fashion, hopelessly vulnerable. You cannot fight the sky. In the end you will hear voices, you will weep, or you will kill, or abandon what you struggled for and flee, but by then the sound of wind will always follow you. Even the great passages on the trail were better. Behind the oxen or in the wagons people dreamed, or they died together. The outlaws who galloped downwind were better off. They were moving, but to stay in those soundless vastitudes was often too great for human endurance. Perhaps to avoid such desolation, I had come to town, only belatedly to find the wind still following me.

In 1959, as the culmination of an administrative career of some fifteen years in two institutions, I was appointed provost of the University of Pennsylvania. The word provost has a peculiar history. When my appointment was first announced, Sandy Gallagher, a Scotch anatomist who had spent most of his life in Uganda, wrote me humorously a few months before his death: "What have you become? The mayor of a small Scottish town?"

The relationship between the University of Pennsylvania and the University of Edinburgh extends well into the eighteenth century when Edinburgh was a medical center of eminence. Some of Penn's medical instructors had received their training there and Scotch influence was strong. Well down into the 1930s Penn had no president. Its presiding officer was the provost. Then, as financial problems deepened and the administration grew more complex, a presidency was created and the provost's post was redefined as that of the first officer of education under the president. He was elected in the same manner as the president by a combination of faculty and trustees committees. The president and the provost held the two seats of honor, as they do still today, at such official functions as the graduation ceremonies. I mention these facts because not all

(197)

American universities utilize this Scottish title. It has, of late years, been picked up here and there in American educational circles but has been used in widely different contexts, frequently to denote a purely appointive presidential assistant or other posts of a different nature.

Sometime before this event occurred I had fallen to chatting with an old sociologist colleague at that time holding an administrative post in the Washington bureaucracy. "You know, Loren," he meditated, "just two things happen to an academic. He finds he has administrative gifts and that they pay better, so he becomes an administrator and ceases to write, as I have." He sighed, though not unhappily. "Or," and here he contemplated his drink with at least equal weariness, "you write in your field and you attend meetings and run errands until you are a pundit and are recognized at last as worthy of being president of a learned society. That is the choice. There is no other."

I remembered his remarks as I looked around my venerable office in College Hall, where portraits of ancient provosts stared down upon me. This was it, then. After all, I had never been much of an attender at meetings save when I was attempting to help our graduate students obtain positions. It had not been an easy thing in those days; posts were few. And my hopes to go to Africa directly after the war had been frustrated by the care of the department which I had been hired to rebuild. There had been others, I was to learn, interested in that project.

One afternoon, after I had reluctantly rejected a candidate proposed by another department, the telephone had rung. A voice spoke, one not at the very top of the administration but close enough to be intimidating. It was cold, hard, uncompromising. "If you persist in your refusal to cooperate you will be called to explain yourself before the president, before myself, before—"

"Men beat men . . ." the old phrase surfaced in my head; I'm still the scuffy-shoed lad who went away from here to Kansas ten years ago. He thinks that over the Appalachians we're all Indians still and have to be taught our place. He's waiting for me to break and say "yes sir."

"I'm sorry you feel that way," I said, naming a man now gone. "I don't like it, but I'll come. Name the day."

There was a soft click on the other end of the line. For a day or two I waited. Nothing came, the threat remained oral, untraceable. Whether the president had actually approved it would remain in doubt. It could have been a bluff beneath his notice.

I had a grant, in the year I left Oberlin College, to go to Africa. It was in the days when the first fragmented man-apes were just being discovered. The immediately post-war contention over sites and between personalities did not make arrangements easy, but these difficulties were not the primary reason for my return of the grant unused. I had a charge laid upon me when I accepted the chairmanship of the Department of Anthropology: that I was to hold it for a term of years until the department was rebuilt. Of the men who sat around the conference table—president, provost, dean—my memory tells me that I am the last survivor. The reason that I have taken time to narrate the call that came to me from that powerful official is to explain why the African grant had to be regretfully returned to the foundation which had given it.

Had I gone away from the campus for any length of time, lurking forces would have sought to staff and commit the department to a course I deemed unwise. The responsibility was mine and mine alone. The grant was returned. I settled into a post I was to hold for twelve years. And now I was provost by some mechanism I little understood. I learned one thing, however. There was something very exhausting about the office. One provost who had sat in conference with me when I came to Penn had died of a heart attack. His successor had withdrawn after four years. My immediate predecessor, a well-loved man and a brilliant surgeon, had resigned after three years. It was thus that the post had come to me.

The provost by tradition served as an ex-officio member of every committee of importance in the University. The restless turbulence that was to culminate in the student violence of the sixties was just beginning to rap tentatively and then ever more loudly on the door. I dealt with these committees, not unfairly, I hope, nor with lack of dignity. But one night I recall crying out in desperation to my wife, "I haven't read a book in two years." The books lined the walls of my office and there they stayed, the almost living creatures who had sustained me since I had gone on alone to follow Friday's footprint on the beach of Juan Fernández long ago. At a meeting in Washington I encountered, by chance, my old administrative friend. "Welcome to the club, Eiseley," he said with a glimmer of amusement. "In one more year you'll never catch up in your profession. But then, there are other compensations."

"What?" I asked, bluntly. "Name them."

"You will learn, as you doubtless already have learned, a lot about human nature. You will, if you are a good administrator, have served an abstract entity which will shrug you aside when it chooses and all you may have achieved will be claimed by others or totally forgotten."

"It wasn't for that I took the post," I said.

"Of course not," my friend sympathized, "but that book you published before you took office, *The Immense Journey*—the one the London *Times* commended—tell me, do you find time to write these days?"

"No," I said. "If I lean back in my chair a minute to stare at the ceiling and an idea, mark you, just a little cloud, begins to form up there, somebody is sure to come in and explain that if we don't put up a matching grant for some project in his department, the world will fall apart in twenty-four hours. You bureaucrats here in Washington have played a role in that one." I grinned, but I didn't feel like grinning.

"Well"—the man in Washington laughed—"I am now going west to be a dean. You'll see me no more. Remember, we are all in the same boat. But you," he ventured further, closing one eye and peering at me as through a microscope, "are a freak, you know. A God-damned freak, and life is never going to be easy for you. You like scholarship, but the scholars, some of them, anyhow, are not going to like you because you don't stay in the hole where God supposedly put you. You keep sticking your head out and looking around. In a university that's inadvisable. I'm surprised at you. Do one of two things. Go back to your arrowpoints. Or do what I hear you are doing well. But you have a third problem for which I don't have the answer. You're not a fish, nor a fowl, you're a writer, and so God help you, my friend. I can't tell you where to go."

I think it was then that I began to hear the wind in my ears, blowing across that prairie out of which I had emerged. And the little fire licking my father's letters, among them the one with the lines I would never fully see. It was late, I thought desperately. I was beyond fifty and when I passed that mark I had stood in my office doorway and heard one of the new breed of timeless youths remark with surety, "This is his fiftieth birthday. What has he got to live for?"

The flames around the crumpled paper in my brain flared and went out. "Good luck in your new choice," I said to my departing friend. "I used to know that coast. You can find beautiful agates on the shore. The sea is very heavy."

"Good luck to you, stranger," he repeated jokingly, unaware of that word's far-reaching significance to me. "When you finish with that job of yours, and believe me I do understand your financial problem, ask after five years who remembers what."

"I'll know," I said with sudden decision. "And that will be enough. Ask me then." As it turned out, we never met to compare notes. He is gone now.

A week or so later I received a call and, after some little discussion, the proffer of a research professorship at another institution. The offer entailed no loss of salary. The fact that I had served Penn as provost for only two years would, I knew, create embarrassment and undoubtedly ill feeling in some quarters. There would be those who would say I had failed, there would be those who would scramble for the office—all those who regard the deliberate renunciation of power as weakness of will, as a "fall." Fortunately there were wiser heads in the administration who prevailed. For the first time in its history Pennsylvania created a University Professorship comparable to those few similar chairs existing at Harvard and Columbia, interdisciplinary, freeing their holders of committee work, greatly lightening their teaching loads. I was the first holder of an academic chair whose numbers were subsequently increased. These posts were later named the Benjamin Franklin Professorships, since a local foundation heavily endowed the project.

I had decided, after prolonged deliberation, to stay. Because of human attitudes such a decision is never easy. One of the officials of the institution which had proffered me the outside post told me in essence: "You will be exposed, as an ex-administrator, to covert insult and hatred. It just isn't smart. Don't do it." Looking back I cannot, in justice, say he was entirely wrong, but a big university has one advantage over a college in a small town: it is easier to fade into anonymity, particularly if one is freed of committees. I kept remotely to the University Museum and I took care to absent myself on the west coast for a year while the dust settled. Only one word reached me there. "Eiseley," so the rumor ran, "is sick. He has gone west to die."

Well, perhaps a part of me had died. As the violence of the 1960s multiplied I turned down three presidential approaches and went on writing while the flames of the letter danced in my head. In one breath I suppose I could call it a ten-year ice age, in another the happiest time of my life, because in that great, gloomy half-circle office the little flames warmed and crackled about me. I read as I pleased, and wrote for my own pleasure. A former colleague, whom I had once aided, cursed the Franklin Professorship to my face. "Luxury goods," another phrase, floated up from an unremembered source. Times were changing, there was high mobility everywhere. Once when I had passed through a turnstile at the University Bookstore and proffered my identity card in the purchase of a book, the cashier glanced at it and at me simultaneously.

"That's not your card," she rasped triumphantly. "Loren is a woman's name." "Ma'am," I said, astounded, "when I was a small boy, Loren was a man's name. Lorena was the feminine. That was before movie actresses took it over. You're selling my books here. Look at them, or call the manager."

In muttering retreat, I got my book. At last my anonymity was becoming full-fledged. Even IBM cards gave me strange instructions or tried to charge me hundreds of dollars in late fees for library books. It seemed I was slowly subsiding to student status. But no, there was no use making a personal thing of it. This was merely the giant background noise of the universe— order against disorder as amplified by the machine. But the machine was supposed to enhance order. Instead, it frequently gave functionaries an opportunity either to say firmly, "God has spoken, even if mistakenly," or on other occasions to study their nails and remark, "The machine is broken," as though this ended the day. It was tantamount to a priest giving last rites to the dying, glancing upward, and pausing apologetically to say, "God is broken. We'll send a mechanic tomorrow. Too bad, old man, but not my problem, I didn't make Him." A wonderfully disheveled ethic induced by the machine and its anonymous human assistants was slowly arising around me. I wondered if somewhere in computer quarters there would eventually arise a course on the ethics of the machine. It was wellnigh time.

Such were some of my thoughts and experiences at the end of the provostship, but there was no place in them for the feelings of rejection about which my well-meaning adviser had warned me. I was part of a school for which I had come to have an affection; it had honored and nurtured me. The rest was small. Not long ago someone asked me the inevitable question, "What did you do when you were provost?"

"Have you ever heard of the spectral war?" I countered.

"No," he said, looking at me in a puzzled fashion.

"Come along," I said. At the corner of Thirty-fourth and Walnut I paused. "You see," I pointed out, "how that block projects in front of Bennett Hall? The cab drivers don't like it, the truck drivers don't like it, but I like it, and I persuaded the planning committee to restructure it that way. It was a move in the spectral war."

"But why?" persisted my questioner. "You yourself say it isn't liked."

"I'll tell you," I said, "and then you'll know. It's a gambler's war. We all lose eventually. When I first came here my office was in Bennett Hall. I crossed this street every day. The corner allowed trucks and cabs to cut a short left turn here. The street enters a highway that shouldn't run through the campus. Students walk across it carelessly. I have seen at least five serious accidents here and one woman cut down by a truck coming too fast around that curve. Then I became provost and a member of all committees. I made a point of attending the particular meeting when changes on that street were being debated. I went, and I gave eyewitness testimony that convinced them. It's all forgotten now, of course. The extended corner is just there and drivers have to turn a square corner. Slow. No more cut-ins."

"And this is all you remember of the provostship?" asked my acquaintance.

"This and some decent people," I added. "It's really quite enough. You see, I figure that, as on a green table, if the dice are fair, there are some lucky throws. That old short corner here at Walnut gave death a better edge, a percentage. I like to think, though I don't know them, that every year there are people left alive because of what I did about that corner. That's what I call the spectral war. It's unseen, but it's everywhere. If you are going to play, you need luck, sure, but the game has to have a kind of justice to it. That," I added, as we strolled away, "is something I learned from a man while I was provost at Pennsylvania. It struck me particularly, you see, because he was a man so rich you wouldn't think he would be paying attention to little things like that. It makes the difference sometimes in the way you feel about a place. Take this corner—no one always wins here, but some do because we changed the house rules."

The child in the ruined house came back to me. "It's all I really care to remember, do you believe that? And nobody knows at all, not anybody. That's history for you. Who records it at street corners? Only one thing knows," I said suddenly, feeling the wind and how it blew there in the desert where I had fought silently for life, "the Player, and he plays on all the corners of the world. Watching the percentages. But you can inch him over now and then. Two years of my life and that's all I remember? No, I remember the face of the woman lying by the truck and the little purse still in her hand. Now this corner will never make that accident as likely again." A kind of rage came over me, but I suppressed it.

Power, this was what it was for, not the humiliation of men. We had saved an old frightened horse on that corner, too. He had almost broken a leg and had been led up on the campus, shivering and frightened, to be comforted. Let the winds blow endlessly through those lost farmsteads from which I came. What if our wars were spectral? We knew for what we fought. Life, life for the purposes of life, and is that then so small?

CHAPTER 20

The Palmist

"YOU will die by water," said the palmist with the shrewd but shifty eyes. That was a long time ago in a dingy Los Angeles flat in the time when I was ill. I had lifted up my calloused palms and gazed at their ingrained soot and iron rust. "I almost have already," I said. "I was named for a drowned man."

Perversely the memory came back to me there under the sun umbrella on the terrace of that Barbados hotel. The man opposite me was one of the new lords of the computer. I had missed the beginning of his story because, for the first time in years, that grubby prophecy had swum before my eyes as I had looked far out across the sparkling waters of the Gulf.

"There are three rings of memory," he was saying, "and the young man was so badly injured by the fencing foil that now he exists in the first ring alone: short-term memory. He is polite and gracious in your presence. When you leave the room, you cease to exist."

"Then he will be always young," picked up the third man at the table. "Life, events, will be a constant surprise. He will not sink slowly into age with an encrusted memory like a coral reef to torture him. There will be no pain, no regrets, no—" he hesitated.

"Who knows?" said the other darkly. "We are still very ignorant. Some people have total recall and may suffer from inability to select what is important. Their brains may be as
stuffed with the extraneous as the house of one of those misers
who preserves old newspapers and broken machinery. This boy
who has escaped that fate will never truly love, never cherish
anyone, because he lives in the present."

"Perhaps an echo will sometime reach him," I ventured.
"The fact that he mostly forgets does not mean there may not
be a haunted cavern to which he returns inwardly and listens
for a sound that never comes.

"I once knew a retarded man of thirty," I added. "I was a
guest in the home of his family. He followed me about with
maps and other little things because I gave him grave attention. You would think he, too, was living in the first ring like
an animal. Then, one evening by the fire, he said something
that showed the hurt of departure, only he had it all wrong.

"He told me, and there was a sick incomprehension in his
eyes, that the boys he had played with had all gone away, had
vanished. Did I know where they were? He wanted to play with
them. And I realized in one stunning moment that to him it
was like yesterday. He didn't know a quarter-century had passed.
He didn't know the children he had played with were grown
up, had disappeared into the adult world, leaving him to mope
uncomfortably about the house. No, to him they were all still
young together and so was he. Only they were just a little long
coming back. He wanted to be reassured.

"Could I tell a 'boy' like that he wasn't a boy? Of course not.
Relentlessly I drove myself into whatever prison he inhabited.
I looked at the maps he brought me and made up what I could.
He would never again see those children for whom he yearned.
I shuffled and evaded and entertained, and all the time he
looked at me with the bright attentive eyes of a dog, as though
he expected his little friends to come dancing in. Here time

(207)

had also stopped but was expected in some dim way to run at childhood's pace forever.

"There are so many kinds of time," I considered. "The three rings, immediate, intermediate, and long-term, is a good simplification, but what to one individual may be short-term and evanescent can be received into the third circle by another. That is why our so-called memories may be diverse or non-existent even over one day we have all shared in the past."

I was tempted to give an example, but the tide was going out, leaving miles of sand flats stretching off to the horizon. My companions were groping learnedly amidst their professions. I got up, still seeing the palmist's eyes. "I am going for a long walk," I said, "out there on the flats." They nodded absently. I slipped off to don my bathing trunks, not because I was a swimmer, but because it was the way of the islands and the way to walk through the sand flats and their hard ripples filled with tidewater.

I walked forward a long distance, a slight squall spattering the sand with raindrops. Raindrops on tide flats leave fossil evidence of their passing just as footprints do. Discoveries of that sort in the early nineteenth century, like those of the footprints of wading dinosaurs, had proved earth carried her own deep-ring memories of so simple a thing as a ten-million-year-old rain.

I tramped and waded for miles before circling back toward a jetty that serviced a supply boat moored alongside. The sea was beginning to inch in over the flats that would someday be part of the island. I waded farther out into waist-deep water that plucked at me impatiently. Once, I thought I caught a glimpse of a long dark shadow, but then it vanished. A voice hailed me from off the jetty. I swung my head impatiently, trying to pick up the cry. The man swept a hand backward toward the sand.

"Shark," the cry came faint and tiny on the wind. At the same time, a little ahead of me, the bullets of a repeating rifle in very accurate hands began to kick up the water.

I walked steadily backward until I was safe on the sand. Had that been the shadow I thought I had seen in the water? Who knows? From up on the jetty the man might have seen a fin, or a clearer silhouette. Or maybe he was just being careful.

I was oddly unalarmed. It was almost as though I had been summoned only to be interrupted. The shark was safe. No bullet had reached him. He had come in as far as he dared and I had come equally far to meet him. It would have to rest at that. Behind me the tide was now sifting sand over the marks of raindrops and a strange footprint, my own, wandering along the shore. I was no different, really, in the memory of the world from those great lizards whose footprints had long since hardened into stone. Back there on the shifting tideflat was my little claim to immortality—earth's record of a summer's day before the fifth great ice ground man's cities into powder.

A little restless sigh escaped me. Like the sea of which it was a detached part, my body, in the deep interior of its salt and winding rivers, was already planning something else. It was tired of me. Somewhere within, it was busy washing away or burying memories. It would continue until nothing was left. A long murmurous incoming wave pursued me shoreward and washed around my ankles. I was trudging heavily now, as the last giant reptile might have moved his ponderous bulk in so much sand. I wanted to lie down and rest. This was earth's way with the first ring of immediate memory. Suppress what lived in it. Dissect all that moves to footprints and bone. Wash it over with sand. Like a robot I limped out of the water to my cabin on the shore. For just a moment in a flash of devastating imagery I could see every footprint I had made along those miles of beach.

In the evening there was dancing on the terrace, and songs, songs for the guests, the marimba beat of starlight music, the velvet tropic night. Someone was singing a wistful ballad. I sat in a white jacket by the secluded bar and looked surreptitiously at my hands enclosing the glass. The iron rust and the calluses were gone.

Lightning flared briefly along the horizon. It seemed to illuminate once more the naked footprints trailing back across the tide flats. Immediate memory? No, it was mixed with the long-term memory of a boy trying to pursue another naked footprint, the print of a man named Friday, through a book called *Robinson Crusoe.*

> "There was a gay lass lived down by the shore
> And a sailorman came to her open door,
> But the glass, the glass was falling."

The singer's voice rose into consciousness. It sounded a long way off, as at the end of the string of footprints on the lightning-lit sand. My hand tightened on the glass.

> "They won't be seen any more, any more,
> The man and the lass in the open door,
> For the weatherglass was falling."

I started up from the table; the rain was coming now in sheets. One could hear the rumble of the surf across the balustrade beneath the sea wall.

The voice followed me. Hurricane warnings, the gathering wave, the falling glass. I could see the shrewd pointed nose of the palmist, the face of a weasel, a very bright bloodthirsty weasel. Had she implanted the notion there, studying the hopeless hand of a wanderer? Her face was obliterating everything else in the night. I followed it through the pouring rain. I came out upon the wall. Below me the great engine of the sea was grinding at the island's edge. In a flash of wet light something indistinct and slippery seemed clinging to the descending stones of the sea wall. It lifted a black arm like a scuba diver and gestured toward the hole beneath me. Ah, how many years had I met the variants of that thing? This, this too, was half a dream, and the song, even the song, was allied against me, promised nothing but treacherous weather, as the earth promised nothing, as life promised nothing but another day.

I locked my hands on the iron rail of the balustrade and shook my head, to and fro, to clear it. Weasel face was gone. I even leaned over the sea wall, as in all those years behind me, to beckon up the black glistening shape for combat. He was gone, no longer pushing at the doorway of my mind. The great engine still pounded and slobbered at the stones.

"There are the three rings, that is all," the engineer was saying from somewhere far behind me. Was he, too, part of the plot to confine me to a narrowing circle for the convenience of my invisible antagonist? Well, he would not. I was running desperately around the circle of long-term memory, but something short-term, something immediate, another ring like a quantum jump, was in the secret. My mind was leaping back and forth between the circles as desperately as a fleeing fox in a maze of hedges. I walked slowly and deliberately back to my cabin. Far off, the singing followed me. It was on another plane now. Its power was broken, but the palmist, of her I am less sure, for there was a sequel, albeit in a different context.

I had been for several days the guest of the director of a large private library in which I had been pursuing some historical research. I had lingered till near midnight and, with permission, would be one of the last to leave the building. I walked down an echoing corridor through the long silent tiers of stacks that contained all the madness, all the wisdom, all the loneliness of centuries. Here imperishable thought lay waiting in the great social brain, waiting to strike fire in minds of similar affinities. It was outside the ring of mortal memory. The men who had recorded these thoughts were mostly dead.

Some of the books had been unread for centuries and yet still sought to be lifted up, perhaps in an era gentler than our own. I wished I knew the destiny of this brain in which I was but a momentary, ephemeral visitant. I had been an archaeologist and I knew too well what fate devises. Everything drifts by fire and flood and ruin into the final ambiguous lettering of the earth's own book of stone.

The dust in these corridors was falling invisibly across the moonbeams from the slit windows as I walked. That dust, I began to feel, was slowly rising about my ankles, choking me as in the sands of a ruined pueblo. I had come a far way in the night and must go home. The constellations gleamed in altered circles in the window slits. The brain lay shattered now, but its essence still smoldered among hanging spiders and fallen roof beams. It could still be rescued if the diggers would come. I poked at a dust-dry binding.

Another stone dropped from the roof. Everything was now shrinking, contracting, the air damp with mold. Though we were inland, I had an invisible sense that the sea was closer, the coastline foundering somewhere nearby. It is a bad location for a world brain, I thought briefly. The architects had not cast forward far enough. I remembered that day on the sandspit and the way the shark came in through the waist-deep water.

Perhaps I had taken it all too literally. Perhaps, because my own books lay here, the woman had seen through me into a vaster ruin than the body of one man. There was a dripping about me. If I waited much longer the molluscs and the coral would come. It was for this that the old seeress had probed far into the pulsing layers of my brain. Well, it was now midnight. I would go back, even if my own words on paper lay slimed and unreadable beneath the tracks of sea snails.

I did go back. Midnight makes all things possible. Tier by tier, stack by stack, I paced up the long corridor. The roof still stood. A faint familiar humming resumed. Insensibly the stacks began to transform themselves to those vast rooms of life that I had guarded in my youth—the rooms where the tricked and maimed had entered life to strive against the strong and vanish. The humming grew—the murmur of many voices contending murderously or gently. The Brain was talking to itself, carrying on some vast dialogue I was incapable of deciphering, though once I thought I detected the sound of my own voice. The Brain was oblivious of its ending in the foundering dark.

I was back where the elevator light showed red. The claws of an old dog began to click beside me. I could feel his low-hung head and heavy breathing. He bore it hard but would stay to the end. In my mind there was a sudden glimpse over a field of ruined tenements at evening and a patiently stroking tongue. The door sprang open into the empty foyer and the night guard touched his cap.

"I used to do something like this myself once," I said to him. "A bit lonely, isn't it?"

"I make out, sir," he grinned. "Doesn't pay to see or hear too much now, does it?"

"Right," I said politely and walked toward the outer door. I held it open for the patient clicking that still followed. Then I was alone. It was a cold night with starshine and I had been far forward to the end. The palmist had shown me the true death by water, but a sweetness inexplicably lingered.

In the all-night bar at the corner I asked for a shot glass of bourbon. When the barkeeper had turned, I made a faint gesture to the face before me in the mirror. "To whatever I won with the dice in childhood," I murmured. "And to the last cast." As I sipped, the old expected tremor in my hands was stilled. The palmist was gone.

PART THREE

DAYS OF
A DOUBTER

The Alchemical meditato is an inner dialogue
with someone who is invisible, as also with God,
or with oneself, or with one's good angel.
                    —Ruland the Lexicographer

CHAPTER 21

The Blue Worm

A BIOGRAPHY is always constructed from ruins but, as any archaeologist will tell you, there is never the means to unearth all the rooms, or follow the buried roads, or dig into every cistern for treasure. You try to see what the ruin meant to whoever inhabited it and, if you are lucky, you see a little way backward into time.

When one hunts for man as I have done, even dead men and their ruins, one goes up, high up into mountains where they may have fled and built in some final extremity, as at Machu Picchu, or down into deep arroyos where their bones may protrude from the walls, or their mineralized jaws gape in the gravel fans. Or one enters caves and with luck comes out again, but not necessarily with treasure.

After so many years of this pursuit the names begin to falter on one's tongue. The Sierra Madres blur to the Sangre de Cristo range, or the Guadelupes west of Carlsbad. The roads, not those fine asphalts or paved ones of today, went up and up and twisted and turned like life itself. Also, and this is important, there were no guard rails then.

It was on such a road, in such a year, that I rode with Manuel. He was what we now speak of politely as a Latin-American. In those days other words were more openly used. These made for frictions and I was never guilty of such usage. I called him Manuel and sometimes *compadre*, but on this particular day it so happened that we were snaking up a thousand-foot cliff when there arose between us a thing called *machismo* and which, though I know little of these matters, is now proudly spoken of as *macho*. You have, or you have not *cojones*, friend, balls, in short, guts. And to prevail over the *norte americano* in these matters is to establish face. Guts, do I make myself clear? Just why we were climbing this trail that might have been blazed by Coronado, and trying to do it in a Model T Ford, I can no longer remember.

Suddenly out of nowhere Manuel said: "*La vida*, life she no matter, eh?" His English was really very good, but the place was not suitable for such remarks. I studied his face a moment. It was brown, open, his grin infectious.

"*Quien sabe?* It could be," I ventured, and looked down a sheer five hundred feet and a tumbled rock slide for a thousand more beneath the cliff. My side, not his.

"You are *americano*, no? *Estados Unidos. Vida* is much to you, to all *americanos*, *mucha vida*, too much. But, how you say, *muerte*, death is the end of all that." He was a fine, reckless driver who loved to philosophize in such places. Particularly with his hands not on the wheel. I watched him covertly. He was doing it now, and he knew I knew it. *Machismo*. It is like bullfighting. The principle is the same. If I stepped out of the car I was not a true man of *cojones*. This didn't bother me, being considerably older than Manuel, but the truth was he had played it too fine. There was just nowhere to step down without some loss of vital equipment that went beyond that little trifle of *machismo*.

I was careful to be indifferent. A way had to be left open. Manuel was a very reckless man. But to be indifferent is also *machismo*. "*Quien sabe*," I shrugged. "The priests, doubtless, would know." Very carefully and slowly I drew a clasp knife from my pocket and began paring my nails.

THE BLUE WORM

"Who knows what matters, Manuel? The good God above. *La vida y el muerte,* they are equal, no? First one, then the other."

To tell the truth, life interested me more at that moment than anyone's *cojones.* I myself was enjoying the view, as would my not-so-sainted mother who had left me the eye, if not the hand, for it.

"Manuel, *mi amigo,*" I reiterated, *"arriba,* up, up, let us try again. It is not good down there. You make your point. You are *muy bravo.* I do not like this hanging about up here, particularly on this side. Life is not likely to matter here. You are *correctemente.* Let us then go where life is. *Vida, pronto. Chigale,* Manuel, *chigale.*"

The engine ground and the car bucked upward. Metaphysics was no longer an issue. A few stones dropped into the abyss. I did not look after them.

But the point is this: *La vida,* she *does* matter; the sound of a gas-flame burner that you never expected to hear again, even the trickle of running water, are both beautiful. And the view, I repeat, *amigos,* is *muy importante,* this view of life. It may not be your view, you need not believe it, but it is mine.

It remains after the violence, after the dissecting laboratories, after the archaeological trenches and the naked, disclosed bone. All there is, *comprende?* And why do I speak here toward the end, resolving nothing, seeing nothing but a dreadful duality of power, the good and the evil. There was not time for everything, *señores.* One must not be boring, but the essence is here. Here are the sounds my father, the actor, heard and spoke beautifully. Here is the beauty my deafened mother could see with her eyes alone and strive to paint. Perhaps you do not care for these things, my friends, but I care and I have come a weary distance. My anatomy lies bare. Read if you wish, or pass on.

The pain has all been suffered and men can do no further harm. Is it not so among you, as among us, that one will think of a woman and grow still, or one will finger a knife and think of wrongs endured, but close it quietly again because the deed happened long ago and one cannot cross the generations? There is no bridge, *señores*. Know that there is no bridge, else I and all of you would hasten to cross and our numbers would splinter the planks beneath our feet. This is not for the ears of *el papa del mundo* but it is true. The knife is still in my pocket, *amigos*. The blade is green with age but I carry it still.

I tried awkwardly to say this to Manuel and his friends by a night fire on the mountain, gesturing as to my deafened mother long ago. They nodded politely. The knife, they understood; *muerte*, they understood; the bridge, I think they understood; but what was to follow I do not think they knew, nor did I. I arose in the dawn and left them with suitable polite salutations. It was good to be alone, always to be alone. For what I pursued was not *machismo* and they were untutored folk who lived by a code that was different from mine. I went on, and, in the words of John Bunyan, ignorance followed, but there was this difference: I knew the creature. I could, in fact, speak to it at any time I chose, though it would never answer directly. It was not involved with pride. It was merely the effect of standing in the sun, a shadow behind the truth. Some men do not like such companions and try to exorcise them. A student once defied me because I spoke of human limitations. They were not relevant, he said.

Nights later, in a powerful dream by the campfire, a gigantic hooded figure like that of a monk sat on a log opposite me. He made no move. Still he raised his cowled head at my uneasy stare. Beneath the hood there was no face, nothing, merely a chill like the void. I endured the moment unfrightened. The compulsion to look had come from me, had been projected, as it were, from my own darkness. I had been drawn relentlessly to peer beneath the cowl. So I went on, down many trails and many years, and, ever true to me like a faithful dog, ignorance followed. Now, at last, the journey was approaching an end.

THE BLUE WORM

I stood in summer before a drab apartment house in a drab street, trying, after a visit, to say goodbye to my mother. World War II had passed. Lincoln was no longer a gigantic over-crowded air base. My mother and my widowed aunt, with the perversity which always seemed to characterize their financial activities, had, just prior to the war, disposed of a property in which I would have been perfectly willing to maintain them. Without consultation, tempted by some unscrupulous realtor, they had destroyed their own security.

A succession of shabby rooming houses had followed because the eccentricities of my mother and aging aunt did not make them attractive tenants in a town swarming with students, pi-lots, and technicians, civil and military. My mother's and aunt's last effects were pilfered or forgotten in attics and basements. I sustained their needs but I was far-off and mementos of child-hood and youth that I would gladly have cared for were aban-doned for the strings, hairpins, and peculiar objects that slowly absorb the attention of the aged. Long-valued possessions were given away or dispersed aimlessly even while I had asked help-lessly for them. I had become an abstract source of funds upon which they were dependent but, as a person, I was less real than the janitor who could be bribed by rare old china to go to the grocery store.

Of this particular day I have only a single street memory. It was the last time I was to see my mother alive. She was nearing eighty, but with a strange, grasshopper gaiety I have never seen in another aging person. Once beautiful, a coquettish youthful-ness and a grotesque overuse of rouge marked her features. Yet out of this ruin still peered the perky intelligence of a naughty child. We exchanged some remarks there on the walk. She could not hear my words. I pantomimed.

Suddenly, with a fey, unreal sense of the future which came upon her at rare intervals, she remarked to me, looking at her arm as though seeing it for the first time, "I wonder why that vein sticks out so." Vanity? Perhaps, but it tore my heart. Like

(221)

a small bird, she cocked her head and looked up at me as though we were seeing together for the first time, as though she, in her way as much a wanderer as I, had just at that moment felt the passage of time. "Why," she said, surprised, running a finger down the blue vein of her forearm, "I believe I'm getting old." Again the bright sparrow's eye looked up at me, escaping as always. "Do you think that's it?" she repeated doubtfully. "Old age?"

I shook my head wordlessly and turned away, raising my hand in a combined gesture of despair and farewell. The last I saw was the blue vein creeping down her arm as she repeated in a voice that seemed to emanate from another dimension, "I'm old, I think I'm old."

The thought was contagious. I extended my own arm later on the airfield, while waiting in the blistering sun. By the powers of heredity a blue worm was beginning to inch its way in a precisely similar fashion down my arm. The culmination was still some years away. "Age," my mother had said, as though struck by the thought for the first time through her little artifices of cosmetic paint. "Age." The same blue worm crawled faintly along my forearm. I thought briefly of the lost photographs of my childhood. The resentment faded. What did it matter, I thought; the blue worm took everything in the end. And there were no children. The photographs could fade in forgotten attics. It was just as well.

Each man goes home before he dies. Each man, as I, physically or mentally, it does not matter which, goes shivering up the dark stairs, carrying a taper that sets gigantic shadows reeling in his brain. He pushes through the cobwebs of unopened doors. Or, rich and happy in his memories, he runs swiftly up the steps of a mansion that has no terrors and bursts into the lighted room of peace to find the fire dead upon the grate and the rocking chair still swaying slightly. "Wait, wait," he desperately intreats, but the last spark goes out upon the hearth and a rising wind slams the door behind him in a fury of postponed violence.

My mother died at the age of eighty-six. The woman who in all my remembered life was neurotic, if not psychotic, whose blasted sense of beauty had been expended upon the saloon art of prairie towns, was dead. Her whole paranoid existence from the time of my childhood had been spent in the deliberate distortion and exploitation of the world about her.

Across my brain were scars which had left me walking under the street lamps of unnumbered nights. I had heard her speak words to my father on his deathbed that had left me circling the peripheries of a continent to escape her always constant presence. Because of her, in ways impossible to retrace, I would die childless. Today, with such surety as genetics can offer, I know that the chances I would have run would have been no more than any man's chances, that the mad Shepards whose blood I carried may have had less to do with my mother's condition than her lifelong deafness. But she, and the whisperings in that old Victorian house of my aunt's, had done their work. I would run no gamble with the Shepard line. I would mark their last earthly appearance. Figments of fantasy I know them now to be, but thanks to my mother and her morbid kin they destroyed their own succession in the child who turned away.

For years I had expected to be drawn back to a deathbed scene of violence, without dignity and without even animal restraint. In the end it happened otherwise. Mother, who at the last had appeared the living embodiment of those witches who had terrorized the seventeenth century and among whom her own familial name was enshrined, had skipped town for the last time. She, the center of violence and contention whose ripples were still spreading toward infinity even in the lives of those who had never known her in life, had died peacefully in her sleep. The eye of the storm had passed.

As the sole survivor who could, or would, attend the funeral,

I had come physically home at last, but not in spirit. "The funeral will be private," I wired the undertaker. "Find," I hesitated over the exact wording, "someone to conduct a simple service. My wife and I will come."

On the day appointed I refused to view the body. "I had rather remember her as in life," I told the lady attendant, thinking briefly of the wistful sparrow look and the writhing vein that was now hidden beneath my own sleeve. I was lying politely. In reality I wished to look no more upon that rouged and ravaged face, upon the creature who had brought me into being. Our debts would be canceled only by my own death, and perhaps not then.

I sat resolutely staring at the floor in the little anteroom. I, who had seen the products of violent death by fire and the havoc of air disaster, who had been in my time a burrower among tombs, as well as a wielder of the dissecting knife in anatomical laboratories, had had enough. I would not look at what lay at the end of that corridor. It was as though something was drawing me, something too earthbound to depart. No, I said to myself alone. Power to destroy me you always had. In fact, you almost did. I have survived the indignities and the words that left me homeless. I have survived, and the deeds are done beyond recall, but draw me into that room you shall not.

The music was drawing to a close. The attendant hovered about nervously. I was an unnatural son. I did not weep. I outdid the reserve of a professional undertaker. There was nothing in me. I was empty of gratitude to this woman who had given me life and at the same time had well-nigh destroyed it. There was a last bar of music.

"If I may," I said, getting to my feet hesitantly. "If it is not too late." I must have appeared like a sleepwalker.

"Most people regret it later if they do not," the woman murmured discreetly. I walked on, the pull from the other room being beyond my will.

I came to the place where she lay. I did not stand close. She slept forever now, that life of so much violence, the eyes closed that had looked upon the world through the narrowed slits of long deprivation into a world of utter silence. Yet, in another sense, she had been oddly youthful, had refused to accept time. Except that one glance at the blue vein, and then only briefly. She had not honestly believed. Not really with that jaunty sparrow's eye. If ever an exorcist was needed it was here.

I saw the high, full forehead, the profile that still possessed a strange unrealized beauty. I did not look at the mouth. I had seen it in its last years. I had seen it across my dying father's face, asking harshly about his insurance. Aging, I could feel its lines drawing into my own.

It was only a moment that I stood there. She had had her way, as always, in the end. I turned aside without any expression and rejoined my wife. The little cavalcade leading to the cemetery began. At the grave, with an old and wary eye, as I turned from the brief service, I saw, surreptitiously, the lounging gravediggers waiting in the background. One was young, a cigarette drooping cynically from his mouth. The cold wind of March blew among the stones, gravediggers' weather.

Something terrible, defiant, and relentless had not even bothered to follow to the graveside. It had already gone and I had felt its passage. Nothing, mother, nothing. Time had never meant to you what it meant to others. I, by profession, had assumed its entire burden.

I took my wife's hand and walked away from the gravediggers. "You have come the whole way," I said, a little brokenly—"the only one who knew everything, accepted everything." I lifted my hands and let them fall.

She nodded silently.

"There were times we were alone and far apart," I groped again. "The travel, the long absences, the years, not easy." I held her hand with mine.

"There were—" I could not say any more. A man contains his silences.

"You came the whole way," I tried again. "It was different with us." There was something too deep to explain.

"Yes, Lare," she said softly. "We came the whole way."

I stared blindly at my shoes, thinking of many things. "Then that settles it," I said, as if the world were ever settled. We turned. I heard the first clods as they struck the grave box. An impatient sound, I thought bitterly. The diggers might have waited.

CHAPTER 22

The Talking Cat

I T is in the night that things come back to me. Sometimes it is a dream of a vanished face and I wake with a laboring heart, waiting till that moment and that hope has passed. Then I lie back tightly composed. But there are other dreams, violent, undisciplined, so that one strikes out fiercely in the dark. Or one weeps when there is no cause for weeping. Or one pants with fear when there is no longer tangible cause for fear. Or one does not sleep and the pictures in the brain come and go and cannot be stopped, suppressed, or directed.

This is the beginning of age as all my family have known age: my grandmother Corey, who periodically cried out desperately in her sleep for help but who, upon being awakened, never confided what it was she feared; or my mother, who finally stalked the dark house sleepless at midnight; or my father, who in the great influenza epidemic of 1917–1918 came home unassisted, went to his room, and lay quietly for days without medical attention, only his intelligent eyes roving the ceiling, waiting for which way the dice would fall and not, I believe now, caring overmuch.

I can only put it that this is the human autumn before the snow. It is the individual's last attempt to order the meaning of his life before a spring breaks in the rusted heart and the dreams, the memories, and the elusive chemical domain that contains them fly apart in irreparable ruin. Oncoming age is to me a vast wild autumn country strewn with broken seed-pods, hurrying cloud wrack, abandoned farm machinery, and circling crows. A place where things were begun on too grand a scale to complete. Thomas Hardy speaks somewhere about the deep-graven family countenance leaping from place to place across oblivion. Well, I am that face but here the journey ends. No children watch from the doorway as I write. Nor, if they crowded there, would I have a consistent narrative to bequeath them. I have aimed instead to bespeak, in some fashion, the autumn years of all men. My pages contain what I could still make out of this cloud-swept lawless landscape. The task has been made even more chaotic by the fact that by profession I am a student of broken things. That I tell the tale of a cat does not mean that I have no sentiment for my kind. It only means that the coming of this cat was a kind of triumph against all that is crushing us.

Not everyone receives the same truth or exists in the same realm of understanding. I have written an account of this epi-sode because it involves a message, and there are those without messages who like to receive them through the medium of others. I myself, before this event occurred, had perhaps been moved by the sad genius of Hardy speaking nostalgically of the time when simple country people believed that the oxen knelt in their stalls on Christmas Eve. Like Hardy, I was tired of my own skin, of sterilized apartment living. I was tired of com-mercialized "people gifts," I was tired of engraved cards pro-claiming a season of good will as though it constituted a tempo-rary armistice in human affairs. Nevertheless I went to a Christ-mas Eve party and it was upon my return from that party that I met the talking cat. Let us have no smiles. I was perfectly sober, so much so that I stood and debated what to do. There are hundreds of lost dogs and abandoned cats in the environs of great cities. Because of helplessness one steels oneself against many deplorable sights.

As I stood in the grounds of my apartment house a thin snow was beginning to fall and there was a hint in the chill air of implacable winds and drifts before morning. It was just at that moment that I heard a plaintive cry from under the shrubbery near the door. The cry was that of a cat in distress. I have heard many such cries in the course of my life under circumstances where I had been forced to walk on. What could I do here, I thought grimly, taking another step toward the door. All over the world there are starving homeless people and animals. Pictures from the past floated through my mind so that I swung my head in distress.

I groped for the keys to the door and it was only then that I realized these pains that afflicted me were the result of the eloquence of the unseen cat under the bush. He was not merely saying he was lost and complaining about it. With a perfectly amazing eloquence he was going up and down the scale of animal grievance. If I could not completely make out the words, I could comprehend their gist. This invisible cat was informing me of the nature of the world, of his deliberate abandonment, of his innocence of wrong, and of my duties as a human being. Why would I not respond to him?

It was more than I could bear.

"Remember the regulations about pets," protested my wife. "You're getting involved."

"Not yet," I answered, but I came down to the lawn and approached the bush, murmuring some kind of guilty explanation to its still-invisible occupant. Explanations of why I could do nothing, protests against his protests, explaining that I was not really heartless but that events—

For a moment this dialogue continued. Then the voice in the shrubbery ceased. He won't dare to come out, I thought. My conscience is clear. I've talked to him, I've been decent, but there is no food in my pocket. I have explained— A silence hovered while the creature considered my obstinate protests and looked me over with night-wise eyes. Suddenly a small gray and white shadow appeared on the snow. The cat had made its decision. It ran directly to me and rolled over on its back in a gesture of trust. I dropped to my knees. The cat, a beautiful young male, rolled from one side to the other while I stroked his stomach. He made some further remarks about cold and being hungry. I felt the dust of travel in his fur. He had come far. He also talked about the dependency of cats upon human-kind. He retained faith in them. I shuddered but it was not in me to disillusion him. Besides, I was sustaining the burden of our humanity at Christmas.

"All right, all right," I said, gathering him up into a ball in my arms while he purred with satisfaction.

"But the superintendent—" protested my wife. "He won't stand for it."

"He will for a little while," I muttered. "He's got to—we've got to find this cat a home. Look, he talks. I tell you it's like language. He doesn't just meow. He's got emphasis, tone, rhythm. I've never heard a cat behave that way. It's uncanny."

"Well now," said my wife gently. "Maybe the SPCA—"

"No," I said. "He isn't a thoroughbred. He just knows how to talk and they wouldn't discover that right off in an animal shelter. There are too many lost cats. He'd be put to sleep. There wouldn't be anyone to understand him."

"Oh?" said my wife, discomfited and uneasy. "But who are you going to get to adopt him? Why don't you consult some-one who understands talking cats, as you put it? Why don't you ask our librarian friend who lives in the building? Certainly she must know about talking cats. She once had a kitty."

"A fine idea," I said. "Let's take the cat to her right now."

The animal in my arms murmured something inaudible, but made no protest as we crossed the drive.

"Good evening," I said to our friend at her door. "We have a problem, uh, that is, er, a cat. But you see it's very unusual. It's a talking cat."

"Is it truly?" said the small gentle woman who loved pets and like us could have none. She stood on tiptoe and looked at the creature I held. "A talking cat? And what does he say?"

"If you will allow me," I said, "I will put him down and you can see for yourself." I dropped the cat gently to the floor.

"And he is lost, you say? Then the dear love must have an egg and some milk."

"M'warf," said the cat promptly.

"You see," I said. "He understands English, too."

"We will call him Night Country," said our friend decisively, "after that book of yours. And you did pick him up in the night."

"Yes," I said, "but the name—" The librarian had a flair for the Gothic.

"I have named more cats than you know," she informed me severely. "It's a perfectly good name and we must think of how to find him a home. Names help, nice names. It's like selling a book," she appealed to me.

The cat ventured some remote comment that his name was probably Whitey, but no one could be sure. The voice came from underneath the sofa. Night Country was only a pair of eyes receding in the depths of the furniture.

"The egg, oh yes, we have forgotten the egg," cried our friend, her ear carefully attuned to various comments beneath the sofa.

The cat's language is tonal, I thought suddenly. Like Chinese. Pitch and emphasis and a rich range that the egg and milk soon quieted.

"Three eggs for the poor love," added our friend. "Tomorrow he must have distemper shots and a bath. We must find a vet." She looked critically at me. "We must check in the lost columns, we must avoid any charity that might put him to sleep.

(231)

Meanwhile"—and a touch of brisk conspiracy entered her voice—"he will stay here with me."

"We wouldn't think—" protested my wife.

"Ah now," said our friend, "I know about cats, and the superintendent rarely comes up here. He will be safe for a few days."

On Christmas Eve, I learned later, Night Country had slept in his protector's canopy bed with a pink nose resting on her shoulder.

"The dear one warms Christmas," said our friend, looking up from the floor when we returned on Christmas Day. "He reminds me of—" I detected a tear shining in her eyes and came hastily across to rub the cat's ears. She did not finish.

"I want to keep—"

"None of us can," cautioned my wife gently.

"He *can* speak, you are right, he can," whispered Night Country's defender. "He sits in the window while I have a cup of tea and he talks like home—"

She bowed her head, her eyes hidden. I looked away, knowing she was long widowed.

The week that followed was a nightmare. I carried Night Country in my arms to two separate veterinarians because, as my wife quickly diagnosed, the first proved heartless and incompetent.

"Night Country will hate me now," I said bitterly, retrieving him. He staggered from tranquilizers; his pupils were distended until a frightening darkness was all one could see in his eyes. Still he came back unprotesting into my arms. I smuggled him back to his hideout.

In the days that followed he lay on a cushion in our good friend's high window and talked about his new home. I detected an unjustified confidence in his voice.

There was nothing in the paper. Everyone we approached— and we spent hours on the telephone—already had cats. There was, as I had suspected, a plethora of cats. Also, the number of people we could trust with a talking cat was small.

The time came when someone in our inner circle of acquaintances mentioned the SPCA again.

"No," I said. "Never. They will not understand that Night Country talks. He is just a cat. We have seen already what will happen. A week or two and—"

A strong compulsion took me. I loved this animal from the cold night.

"If it becomes necessary," I spoke my ultimatum to a room where we had lived for a quarter of a century, "if it becomes necessary we shall move, so help me. We won't desert him."

No one said anything. I realized with embarrassment that my voice was shaking.

The next day the phone rang. Our friend was crying but jubilant. One of the secretaries in the department would take Night Country. I breathed a long sigh and held my head against the wall. It was all right then. The secretary was a beautiful, kindhearted girl, although I remembered, a little uneasily, that she was also one of the younger generation that roams from Paris to Milan in the summers. She had her own world. We purchased a little carrier for Night Country to be taken to his new home. He wailed once heartbreakingly as he was driven off.

"Till the spring then," I whispered to myself. "Till the spring."

Months later, at my request, Night Country was brought once to see me at my office. He emerged promptly from the traveling case that I had purchased for him to be taken away in. He was bigger now, but he still rolled over in his little friendly ritual. After this greeting he investigated each box and bone in my office. He discussed them, too, in that strange tonal language that I could no longer follow. Then he sat fascinated on my window ledge and watched a long freight train with brightly painted cars like toys pass on a trestle a block away. His eyes were very wide but he uttered no comment. His new owner had left the room. Night Country was her cat now, though he finally found a little secret shelf among my skulls and peered out inscrutably at me from the night darkness that would always follow him. I would never know if he remembered me.

Not long afterward the girl who had taken Night Country decided to go to Paris and we had to find another home for him. One of my former students who lived in New Orleans was approached. Yes, she and her husband would make room for him, but they had a dog, a big dog. If things worked out—

They did, as it so happened. "He is living in the dog house with the dog," the startling message came back. "Shep took to him immediately. We have never seen anything like it."

Ah, but I know, I thought, tilting back in my chair. Night Country must have talked to Shep and rolled over. He is walking through the world like a cat out of Eden. A Christmas cat. A cat who talks to each person once, just once. Now he has escaped the snow and is living in the house of a companionable great dog. Nothing, no one, has ever broken his trust, since he made his decision and came crying to me from the bush on Christmas Eve.

Where had I read about an old circus lion in Britain who had escaped from his cage? They had found him on the moors bedded down with some sheep he had not harmed. It was the Christmas feel of kind, I thought, for the variegated life of the world across the boundaries of form, the thing so lost to most of us, save for the confident talking cat and the lion and the wistful thinking of poets:

> for I
> love forms beyond my own
> and regret the borders between us.

This was not the Christmas of the engraved cards. It was the message of the talking cat I had rescued from the snow. He had spoken just once and I had understood him. I did not have to hear twice. I would never forget. I had wanted him for myself, but he was a message carrier and messengers cannot think of themselves. They must go on. I thought of him sleeping far away in New Orleans between the paws of a shepherd dog. I thought of the old lion who had slept a few weary hours among the sheep before men came to get him. Then I put on my hat and went restlessly out into the late spring rain of Eastertime. I would never see Night Country again. He was a messenger.

In past years when hunting with others I have failed to call attention to a fox. I have unwrapped a snake from a pheasant and injured neither—widened the eye of the world, in other words, pushed death momentarily aside, but here, this cat, you see, cost me grief and sustained effort. He is still living, may outlive me, and of course does not remember now, but with him I finally outfaced the universe.

We both did it. Let it stand for our steps going away, the voluntary act, trust on his part, response on mine. For a moment we closed the barrier between forms, we talked together. It is not his fault if his brain by now is a drifting haze of unknown faces that did not stay. He commanded me to a duty known between us. Let it stand for the record—I will hold the memory for him.

They called him Night Country and wondered where he had come from. Well, he was a true cat. No one would ever find out. They could only guess from the nature of the message. As for me, I believe I guessed, but I never told. The message can stand by itself.

CHAPTER 23

The Coming of the Giant Wasps

TO pass immediately from cats to wasps is perhaps a strange transition, even for a naturalist, but from childhood on I had attempted in off hours to observe the New World version of the insect realm of Henri Fabre. In fact, insects had fascinated me long before I became an anthropologist. I will never forget that autumn just because of the warm September light on the hillside and the way those great Sphex wasps, the cicada killers, came in over the grass like homing bombers. Perhaps they had hunted briefly about in other summers, but in this year someone had told me moles were tunneling in the backyard of our apartment house. I had gone out to investigate and it had needed only one glance at a quaint, big-lensed, intelligent head coming out of one of the burrows to know that something magnificent had intruded into our little backyard garden.

As the days passed one could hear an occasional cicada's song terminating abruptly in a kind of stifled shriek—a sign that an assassin had reached him. If one waited quietly it would not be long before one of the huge wasps, the largest on the eastern seaboard, would come gliding in toward its excavation with a paralyzed cicada in its clutches. Then the creature would emerge alone, circle at speed to get its bearings, and wing off into the trees.

What I admired about the wasps was the deadly perfection of instinct they exhibited, and their utter indifference to man. If I stood near one of the holes long enough, I might be circled but never attacked. I was merely being utilized as a beacon marker.

These wasps of the genus *Sphecius* are solitary, and giants of their kind. The first one I ever saw, years ago in another state, was carrying prey heavier than herself and getting it airborne by utilizing the corner of a building to bounce her upward thrusting legs against, while using her wings at full power. When she had reached sufficient altitude she had zoomed off in a beeline for her distant burrow. Here, in the backyard, the wasps had selected the sunny hill slope for their nests. This neighborhood activity is their only sign of incipient socialization. They had come to this spot, I suspected, because there was little or no appropriate wild land remaining in the growing suburbs around us. If they took cognizance of human beings at all they would not have intruded here. They frightened old ladies by their mere size, and gardeners trampled remorselessly upon their excavations. But still, in the early morning light, they would return to dig and their energy was boundless.

I was as pleased that autumn as if someone had reported a panther in our pine tree. I was getting old enough to want to rethink what I had learned when I was younger. I believe now that it was the coming of the giant wasps that first aided in implanting some doubts in my mind about the naturalness of nature, or at least nature as she may be interpreted in the laboratory.

These wasps, and their assorted brethren, the tarantula killers, present in miniature several of the greatest problems in the universe. Beautiful as they appear in sunlight, their deeds below ground are less edifying. They would justify Darwin's well-known remark about the horribly cruel works of nature, or even Emerson's observation that there is a crack in everything God has made.

Yet these hymenopterans exercise no moral choice. Their larval stage has to be supported by a peculiarly frightful form of feeding upon paralyzed flesh, while the adults must contribute to their survival by one of the most precise surgical operations known among the lower creatures. Otherwise the species would perish. Beauty and evil, at least by human standards, course together over the autumn grass. The wasps have always impressed me as formations of fast fighter planes impress me—incredible beauty linked to destruction in the heart of man. As a cynical pilot once put it of his charges, "They are nothing but a flying gun platform."

Similarly one could say of *Sphecius speciosus* that it is nothing but a flying hypodermic needle. But how wrong, how terribly wrong, oversimplified, and reductionist would be both observations. Into man's jet fighters, year by year, have gone some of the most elaborate scientific calculations in the world—mathematical equations of the utmost abstract beauty, which, when embodied in reality, result in speeds outrunning the human imagination. Their potential uses of terror lie in the ambivalent nature of man, or perhaps in the ambivalent nature that created man. So the great wasps, the invaders of the autumn grass, carry navigational aids whose complexity remains unexplained and whose surgical intent is comprehended, if at all, only in the dream that lies below all living nature—a dream as tenuous and insubstantial as the shaft of September light through which roams a flying deadly lancet. "My thoughts are not your thoughts," runs the Biblical injunction of Jehovah. "Your ways are not my ways. I make the good. I create evil. I, the Lord, do all these things."

I have come to believe that in the world there is nothing to explain the world. Nothing in nature that can separate the existent from the potential. I start with that. Biological scientists, however, are involved by necessity in the explanation of life.

In the end many are forced into metaphysical positions which reflect their own temperamental bent. There are reductionists like Jacques Loeb, who strove to bring life into the manageable compass of physics and chemistry, or men such as the philosopher Henri Bergson, who attempted to distinguish life as a separate, indefinable principle, the *élan vital.*

Between these extremes we all flounder, choosing to close our eyes to ultimate questions and proceeding, instead, with classification and experiment. Even then our experiments are apt to be colored by what we subconsciously believe or hope. Also there is life within us, a magnificent, irrecoverable good. It could be argued that there must be something a little strange about those who scorn that good, just as those who love its endless manifestations may be accused of some form of submerged worship.

This conflict extends into the last century, intensifying with the discovery of the principles of evolution. One of the last great evolutionary controversies of that century arose, in fact, over the behavior of the solitary wasps whose mysterious habits, so Henri Fabre, the French entomologist, proclaimed, simply did not lend themselves to an explanation by means of the selection of chance Darwinian mutations. To my mind the controversy was never really resolved, only softened and eventually dropped as other more comprehensible discoveries diverted the naturalists of the new century. Nevertheless the world owes a debt to Henri Fabre, who worked all his life in the sandy stretches of southern France. Fabre was too unschooled to accept readily what he was told in other people's books, including Darwin's. Instead he lay under the spell of the elegant French experimentalists who preferred controlled investigation to armchair theorizing.

It is not sufficient to say, therefore, that the schoolmaster Fabre was an anti-Darwinian who saw, in the perfection of instinct, an utter barrier to evolution. Fabre merely chose, on the basis of his field studies, to ask some legitimate and penetrating questions. Darwin himself realized that among the amazing life cycles of insects there were cases difficult to explain on the basis of pure undirected variation. At one point he confessed in *The Origin of Species* that many instincts difficult to explain could be opposed to the theory of natural selection, "cases in which we cannot see how an instinct could possibly have originated" and "in which no intermediate gradations are known to exist." It was this sort of problem, notably the knowledge of their opponents' weaknesses possessed by the Sphex wasps, which had troubled Fabre.

A careful reading of the French entomologist reveals that he was not unaware of variation in the behavior of his wasps. He was, nevertheless, notably impressed by the surgical knowledge manifested by both grub and adult, instincts which seemed to rule out any theory of their origin through the selection of chance variations. The other Darwinian refuge of that time lay in the suggestion that learned behavior, "habit," might precede and prepare the way for the emergence of purely instinctive behavior.

Fabre had expressed doubts, not totally answered to this day, as to how unaided Darwinian natural selection, or, indeed, selected "habit," could produce something that would be of no natural use until the chain of instinctive reflexes led to the survival of the wasp. Such survival could only be effected through a very complicated mosaic of perfected and interlocking behavior distributed between the adult insect and its larval offspring. To paraphrase one modern naturalist, John Crompton, a surgeon does not learn his trade by indiscriminately pursuing and slashing at his potential clients with a sharpened lancet. Neither is it likely that the Sphex wasps acquired their skill through chance behavior which, in the most successful, slowly froze into the rigidities of perfected instinct.

The fearsome operations of these wasps depend upon an uncanny knowledge of the location of the nerve centers of their prey in order to stun, not kill the creature. The larvae, also, must possess an instinctive knowledge of how to eat in order to prolong the life of the paralyzed body which they devour. To complicate matters further, the victim—even a formidable, outsize victim like a tarantula—seems to have some foreknowledge of its helplessness, some fear of which its agile opponent takes absolute confident advantage.

All is arranged in such a manner as to suggest the victim possesses an innate awareness of his own role, but cannot escape it. If, so Fabre muses, pure chance has, through long ages, decreed this relationship between hunter and hunted, why have not the cicada, the cricket, the tarantula, equally evolved a defense against their fate? If we attribute success to natural selection in the case of the wasp, why has not the same force been at work for the victim? Or, on the stage of life does the victim as well as the huntress play a foreordained role?

Whatever forces have been at work in the evolution of the wasp family, it is clear that they have little, if anything, to do with that nineteenth-century cliché about "the effects of habit," which tells us nothing. Fabre was right in that judgment, possibly right even in his fateful admonition, "It is not in chance that we will find the key to such harmonies." "The man grappling with reality," he concludes, "fails to find a serious explanation of anything whatsoever that he sees."

Let us grant that Fabre chose not to explore the evolutionary road. Let us admit that his metaphysical bent lay in another direction. But the attempt of many of the Darwinian circle to explain the mysteries of instinct was not always enlightening. They confused their own Darwinism by choosing the best of both worlds when they argued that chance-acquired habits might sink into the germ plasm. The experimenter on his little patch of poverty-ridden soil at Serignan had toiled long enough to know that the world he investigated provided more mysteries than answers. In his old age he adhered to that conclusion. Perhaps it was his philosophical weakness. Perhaps, on the other hand, it was simple honesty.

THE COMING OF THE GIANT WASPS

The inorganic world out of which life has emerged and into which, in season, it falls back, possesses the latent capacity for endless ramification and diversity. A few chance elements which appear thoroughly stable in their reactions dress up as for a masked ball and go strolling, hunted and hunter together. Their forms alter through the ages. They are shape-shifters, role-changers. Like flying lizards or ancestral men, they run their course and vanish, never to return. The chemicals of which their bodies were composed lie all about us but by no known magic can we return a lost species to life. Life, in fact, is the product of singular and unreturning contingencies of which the inorganic world disclaims knowledge. Only its elements, swept up in the mysterious living vortex, evoke new forms, new habits, and new thoughts.

I am an evolutionist. I believe my great backyard Sphexes have evolved like other creatures. But watching them in the October light as one circles my head in curiosity, I can only repeat my dictum softly: in the world there is nothing to explain the world. Nothing to explain the necessity of life, nothing to explain the hunger of the elements to become life, nothing to explain why the stolid realm of rock and soil and mineral should diversify itself into beauty, terror, and uncertainty. To bring organic novelty into existence, to create pain, injustice, joy, demands more than we can discern in the nature that we analyze so completely. Worship, then, like the Maya, the unknown zero, the procession of the time-bearing gods. The equation that can explain why a mere Sphex wasp contains in its minute head the ganglionic centers of its prey has still to be written. In the world there is nothing below a certain depth that is truly explanatory. It is as if matter dreamed and muttered in its sleep. But why, and for what reason it dreams, there is no evidence.

It is now high autumn. Apples are falling untended and smashing on the stones I have come to call Wasp Alley. The smell is drunken, ciderous. In the growing dark, wasps of many species—vespas, yellow jackets, mud daubers—clamber over the ripe ungathered fruit. On this particular evening something more formidable rises and bumps my nose inquisitively before it flies away over the roofs. It is one of the giant Sphexes caught in an innocent moment of adult feeding, the deadly needle sheathed at last. Instinctively I know this will be our final encounter of the haunted year.

But still, not quite. The sun, a week later, falls in gold October splendor over the little hillside. Coming home in the afternoon I sit down, a little stiffly, and survey the drowsy slope where the closed burrows of the sphecoids lie hidden in the autumn grass. At the bottom of each burrow reposes a mummy case, a sleeping pupa. It will lie there still drowsing under the winter snows, and surrounded by the emptied husks of its feeding.

Beneath the midsummer sunlight of another year a molecular alarm will sound in the coffin at rest in that silent chamber; the sarcophagus will split. In the depths of the tomb a great yellow and black Sphex will appear. The clock in its body will tell it to hasten, hasten up the passage to the surface.

On that brief journey the wasp may well trip over the body of its own true mother—if this was her last burrow—a tomb for life and a tomb for death. Here the generations do not recognize each other; it remains only to tear open the doorway and rush upward into the sun. The dead past, its husks, its withered wings are cast aside, scrambled over, in the frantic moment of resurrection.

The tomb has burst. A tiny chain of genes and releaser genes in the black dark has informed the great winged creature of her destiny, the unseen flowers, the shrilling of cicadas in the sun. She carries, not alone the surgical instrument, but the map of operations yet to be performed on an insect she has never seen. She is a nectar feeder, but it is for carnivorous grubs that she will labor, the grubs of which she was once one, feeding on paralyzed flesh in the sightless gloom of a walled chamber.

Briefly I recalled the days at the hatchery: one-legged chicks, scissor-billed chicks that could not peck properly—the dreary cheeping orphans whose bodily instructions had gone awry. I knew also of similar human wrecks immured from sight in institutions, or hidden in shamed households where, as a guest, waking in the night, one could hear them cry out in desperation far overhead in an attic room.

"The injustice, the injustice—" a great scholar had once breathed to me, over my account of the hatchery and its horrible methods of disposal. Here beneath the leaves on the autumn grass slept nature, or a part of nature, so beautifully, so exquisitely contrived that it was hard to imagine error, hard to conceive of all the pieces of that intricate puzzle being put together from the blind play of natural selection alone. Looked at from one point of view, nature had created monstrous evil, the tormenting of helpless, paralyzed flesh. Looked at in another way, the eternal storm maintained its balance.

I remembered how that formidable autumn creature had hovered before my face as though questioning my own existence in the apple-strewn twilight. Apples were still falling untended, while far away, on another part of the planet, people died of hunger. The great Sphex itself was doomed in the oncoming frosts of autumn. Everything living was falling, disintegrating as under the violence of an unseen hurricane.

"Created to no purpose by an endlessly revised genetic alphabet," one part of my mind contended. "A work of ecstasy," the words of Emerson echoed in another chamber of my thought. "But the injustice," pleaded my grand old scholar. The leaves continued to fall silently. Why should I ally myself against his protest on a day in which my own bones were stiffening in the autumn sun? Who was I to rule against that judgment?

Throughout September I had watched the tiger-faced Sphexes

digging with furious energy. I had heard the muffled shriek that ended the cicada's song. On that lonely backyard slope it had somehow pleased me that the wasps came and went as though I belonged to another world they chose to ignore, a misty world for which they carried no instruction, just as I carried none for the totality of the night. Though shorn of knowledge, willing to accept the dreadful otherness of the Biblical challenge, "your ways are not my ways," I had come to feel at last that the human version of evolutionary events was perhaps too simplistic for belief.

There is a persistent adage in science that one must not multiply hypotheses unduly and without reason. I grant its usefulness. Nevertheless it can sometimes lead to the assumption that science finds nature simple and that someday all will be known. Vain delusion, incredible folly, I thought, brooding there at sundown over the sleeping surgeons known as Sphex. We, our species, will be gone before we know.

I drew my stiffened foot beneath me. As in the case of the French observer on his little sand plot, uncounted mysteries had a way of persistently intruding into my mind. The wasps' master chart of surgery was not always perfect. Still it was terrifying enough to provoke the envy of any practicing physician. This evolutionary marvel was just not that of slow selection for size or greater running speed, as among horses. The entire pattern had to work or the species would perish. There seemed to be no intermediate possibility. The larvae have to feed in a certain way. The adult female has to seek prey upon which it has never personally fed. It has, furthermore, to identify that prey. The wasp has to bring its paralyzed cicada back over a distance to a burrow it has already constructed and whose position it has previously mapped with the care of an aerial navigator. To explain this uncanny phenomenon by computerized armchair genetics may be theoretically possible if one starts from certain current assumptions which leave me vaguely uncomfortable. Perhaps that can be termed my metaphysical position. I am simply baffled. I know these creatures have been shaped in the cellars of time. It is the method that troubles me.

Some ten years after Fabre's death in 1915 Alexander Petrunkevitch, the spider specialist, had described his own adventures with a tarantula-killing wasp, *Pepsis marginata*. All of the great wasps are fascinating in their diverse surgical habits, but what had long intrigued me about this particular account of the Caribbean killer wasp, *Pepsis*, was something that seemed once more to lie doubly out of time and belief. Shifting my foot again in the misty light of the Sphex graves, I tried to recall it. It was important now; I had not many more autumns in which to ponder such problems. *Pepsis*, the tarantula killer, dueled with a far more formidable creature than a cicada. Its knowledge of its prey's anatomy was just as deadly as that of the giant Sphex. But one thing more, one bit of preternatural intelligence, continued to challenge my faith in the pure undiluted chances of natural selection.

When *Pepsis* paralyzed her giant foe and deposited her egg, she added one more complicated pattern to the behavior of the killer wasps. She packed her big, hairy opponent so masterfully into its grave that it could never dig its way out even if it were, by some chance, to recover. Every limb of the huge spider was literally handcuffed to earth. Poison needle, utter paralysis, were not enough. A final act of devastating ingenuity had been added.

The autumn light was growing dull about me, the shadows were gathering. I was beyond the country of common belief; that would seem to be the source of my problem. I had spent a lifetime exploring questions for which I no longer pretended to have answers, or to fully accept the answers of others. I was slowly growing as insubstantial as the sunlight on this hillside. I could not account for myself any more than I could validate in material terms the strange anatomical charts that slept, for now inactivated, in the tombs beneath my feet.

Slowly, painfully, I arose and limped away. As I walked I knew, with the chill of a not too welcome discovery, that I was leaving the sharply defined country of youth and scientific certitude. I was seeking an undiscoverable place, glimpsed long ago by the poet Shelley

> built beyond mortal thought
> far in the unapparent.

Strangely, in a little-known passage the great experimentalist Claude Bernard once echoed, more grimly, the same idea. "I put up with ignorance," he said. "That is my philosophy." Thus ended the visitation of the giant wasps. I never saw them again.

CHAPTER 24

The Time Traders

CRITICS who have studied the records of men's lives have been known to observe that the best annals of life are those in which a man strove against adversity and won, or, if not that, left an account of how his character was shaped. Of the record that the reader has followed thus far, it may be said that I appear to know nothing of what I truly am: gambler, scholar, or fugitive. I have played a certain visible role in life, but that my thoughts have often been elsewhere is quite apparent. I seem preoccupied with chance, whether it be the chance that determines life or death upon a street corner, or what it may have been that hovered about me in the ruined farmhouse where, as a child, I threw dice, mimicking a game whose scores I could never possibly determine.

Toward the end, called variously writer, naturalist, scientist, I feel impelled to deny everything and hide what is left as best I may. Perhaps the geologists have a better name for my predicament. They speak of outwash fans, and such a fan is very similar to the autumn of our lives. The plunge of pebbles in a rapid stream, caroming against each other, has ceased. The last valley dispersion cannot be carried further. Entropy has conquered movement, whether among the stones or the episodes of a life.

Just so the battered and shaped remnants of dreams and realities lie quiescent in the aging mind to be strolled over, picked up, or dropped again indifferently. The white or black stones of experience are not meaningful because there is no longer an onward impulse. And if we can find none, do we collapse finally into a juxtaposed if invisible spread of memories, each one representing a stone that might have deflected us elsewhere?

In reality, I am not of this persuasion. We are not to be found among the stones, we have been the stream. And it is the stream, not the colliding boulders, that makes up a life. Without the torrent the boulders do not clash, nothing moves or is bound anywhere. The stream, if we may pursue the metaphor, is the life energy that sets events to reeling and colliding. If the boulders are big enough they may momentarily impede the stream, but the stream, life, is the energizing power. Events are its creation. I merely observe, therefore, that as a child, the dice thrower, I was setting events in motion just as much as when I delivered the speech of the century to one drunken listener, Nelson Goodcrown, under a guttering street lamp in Tia Juana. The stream was flowing downgrade following its temperamental bent, its own peculiar gravity. So was that unutterably powerful torrent, Nelson Goodcrown, as we fenced with words over the whores of that border town.

Then our two energies raced away to other destinies, mine to caverns, lost fragments of history, broken cities, strange animals. Headlong I pursued such things, tumbled them over, or circled them warily, trying to understand. In this attempt to understand, to become civilized, to be reflective, one becomes by turns gambler, fugitive, and scholar. What eventually lies on the outwash fan is memory, and it is from memory that we hesitantly try to reconstruct the nature of each individual torrent. Our energies are fierce, but unlike water we possess a power to flow toward the circumstances that create our final destiny. Thus we are nearing the status of the inert valley floor. In what has come to pass, it is for the reader to detect his own gambler, himself as fugitive, his own rebellious scholar. In the end it may be he will have discovered personal secrets and in the resulting confusion I will have achieved my purpose and effected once more my own escape. The stream has entered a wide valley but a faint persisting energy still plucks among the stones.

I was made aware of this when circumstances led me once more to an indirect encounter with trade rats similar to those that had stolen my glasses and escaped with other sundry belongings around the cabin assigned to me in the desert long ago. The genus to which my little persecutors belonged was eloquently described by the naturalist Ivan Sanderson: "These animals collect great masses of sticks and other material to build huge nests in caves, at the base of trees or in branches . . . but why they should pilfer tin cans, jewelry, bits of glass and all manner of objects, and often replace them with other items . . . is quite beyond man's understanding."

The general argument runs that the animals are picking up objects for the huge dens in which they encapsulate themselves and that they merely drop one item for another. Why their tastes should run to wrist watches, glasses, and padlocks is anyone's guess. Remembering their midnight scamperings, I sometimes wonder whether they also capture the intangible and bear it away into attics and mountain crevices. Old voices, perhaps. I think wistfully of a wanderer whom I had once picked up on the road and who had stayed a day or two in my cabin entertaining me with beautiful baritone renderings of ballads from the border dives. Does *La Cucaracha* still sound on a quiet evening in the cabin attic—faint now, far-off, listen, listen, a half century has almost gone, but the rats might have kept the tune—they, if anyone.

Why should I think of the pack rats? I had not seen one for many years and with the encroachments of civilization they are growing sparse except in waste places like the desert and the mountains. Archaeologists and biologists are concerned with investigating the climate and flora of the west, from Canada south into the deserts of Mexico. There are plant remains that tell the story of climate changes which extend over forty thousand years into the time of the ice age. If one can study their distribution and date them by the well-known carbon 14 technique, one may have clues bearing on plant, animal, and human movements. It is the rats who keep the record.

In the early 1960s Dr. Philip Wells of the University of Kansas began to find out. These bits and pieces of the climatic past, just like beer bottle caps and old padlocks, were hidden in the nests of pack rats. Some of their more secure dens have been inhabited by the little pawnbrokers for thousands of years. Successive generations have toiled at their huge middens. Since the rats are not travelers, vegetational remains associated with their nests mark the changing weather of ages. The rats at some time in past millennia have secured the plants nearby. Because woody remains can be closely dated by the carbon method, enormous progress has been made in studying the postglacial vegetation of the plains. The necessary material has come from pack-rat middens; they are calendars.

Eastern Nebraska today is a no-man's land on the edge of these scientific developments, most of which have been carried on in the more arid regions bordering the mountains. In other words, long-time pack-rat lairs are sparse, and the animal does not flourish well in too close contact with man, however attractive he may find the belongings of prospectors and campers. On the other hand, there are still uncultivated spots and exceptional locations into which the wandering rats can venture.

A few years ago I chanced to visit an old haunt near which the fleeing convicts from the Nebraska penitentiary had briefly sheltered on the first night of their escape in 1912. I knew it well from childhood. From where I stood I looked down into a gully carved in Dakota sandstone.

The last time I had visited that spot as a youngster, cotton-wood trees had leaned secretively over a huge stone basin full of

(251)

water. There had been frogs and tadpoles—even a mud turtle
or two. Long ago I had played there alone and carved my initials in the friable stone. Now I stood looking into the basin.
The trees were gone, the water was gone, and there was no
discernible life in the basin. The initials had vanished, though
even in childhood I had had the sense to carve them deep
against the encroaching years. The sun beat down mercilessly
upon my head.

A little way up the path by which I had come a farmer leaned
on the fence watching me curiously. I walked up to him.

"Hot," I said conversationally. "Whatever became of the
water and the trees?"

"What water, what trees?" he grimaced. "You don't know
this place, fella. Ain't nothin' like that around here. Never
been.

"Water," he snorted contemptuously and waved his hand at
the sun. "You must be new around here."

"No," I started to say. "I was born—" But then I did not
finish. I saw by his eyes it was useless and that I was standing
among trees that only my own mind told me had once existed.
There was a veil beginning to form in the sun between us. Before long I would be as invisible as the little boy whose initials
could no longer be seen.

"Maybe so," I said and walked like a ghost back into the past.
Two world wars had come and gone since I had visited that
outcropping of Dakota stone. The farmer was younger than I
—young enough to believe in his own time and its appointed
surroundings. There was only a pile of rusted cans where the
pool had been. Pausing a moment I peered over a declivity
and thought I detected slight signs of a midden where some
cans had been dragged closer together. There were a few piled
sticks and a weathered cartridge case or two. And there was a
characteristic stain of rat urine on the rock. "Hanging on," I
thought, "but likely he won't make it." I eyed the cartridge
cases curiously because they brought something back to mind.

I turned and retraced my steps, conscious that the suspicious farmer was still watching me.

When I reached my hotel I threw myself wearily on the bed. I did not want to remember, but I was remembering. I was haunted by my father's face the evening he had been mentally out in that March blizzard in which streetcars had been forced to halt, and sleighs had overturned.

The next morning I visited the library archives. Staring into the microfilm viewer brought the whole episode back most vividly, the grey and black specks flickering past as I sought for the proper month that had cost men's lives and rocked a complacent state. In my haste I was running the year through the viewer like driving snow.

March was the month, a bitter, blizzardy March. Suddenly I was approaching the time. I slowed the viewer. It was the second week in March, the tenth day, to be exact. The warden had perhaps four days left to live. I scanned the viewer carefully. On that day the governor had told a group of prison reformers he did not wish to listen to "generalities" reflecting on the prison management. He also elaborated his views on "wholesome discipline." Apparently he had overruled the protests of his own board who had demanded the warden's resignation some time before. There was a "conspiracy," the governor claimed, to prevent his choice from being seated.

Then the chaplain of the prison, whose conscience must have troubled him, asked for an appointment with the governor to express his concern over conditions within the institution. Immediately his resignation was demanded by the governor. "If you don't like the situation, you can quit," roared that eminent official. "There's the door." The young chaplain resigned; a volunteer women's teaching group was also refused recognition. A medieval gloom hung over the grey fortress, in which at that time the warden had his family quarters.

Was it true—a rumor I remember hearing circulated—that someone in the warden's circle, hearing the screams of tortured men, had once telephoned the governor in protest? If so, it was all useless. "Wholesome discipline" was seemingly in order.

What was this singularly beneficial medicine, I wondered, my mind running back to the sad and carefully phrased words of my father on that fateful evening after the break-out. "Someday you will understand." I searched in memory for later reminiscences. I looked at other Nebraska papers opposed to the governor's policies and the warden's prison regime.

Finally, I began to realize that nineteenth-century prison management had not then died out in my own state, nor in many others. A respected judge, one of those critical of the conduct of the prison, spoke of poor-quality, underpaid guards, of hangings by the hands, the water torture by fire hose, punishment cells whose nature it was easy to imagine. The screams could be understood now, just as I could understand that the ringleader of the break-out was reputed to have whispered to the warden's face, "I'll get you, warden. I'll kill you if it's the last thing I ever do."

I shall not bother with the full story. A few days after the resignation of the chaplain, in the midst of a snow so blinding that not even railroads were running, three men broke out of the pen. They did not go over the wall. One, "Shorty Gray," a long-timer, had been sentenced for bank robbery in Aurora, a town I was later to live in, as though our paths, a generation apart, were destined to cross and recross throughout eternity.

Armed with a smuggled pistol in the hands of Gray, the prisoners rushed the front gate. The deputy warden drew and died without firing. The warden perished at his desk. Gray had kept his word. There remained the gate. A guard died. The skilled user of safe-cracking techniques blew the locks with smuggled nitroglycerin, hastily applied. The convicts were out and running through the storm, thin clothing, a few cartridges, a pistol, and no money.

Gray had had a loyal friend, a Kansas City woman, who vanished from a local rooming house the day before the break-out.

She was never apprehended. In less than a week, with the exception of one man who surrendered, they were all dead. It was thought the woman had brought the gun and the explosive, which had been smuggled in by intermediaries. The survivor never talked, saying Gray, the dead man, was the only one who knew the details. He may have been telling the truth. He was the youngest, and when the posse came and his companions died, perhaps, as the youngest, life had seemed sweeter to him than to the others. I rather think he may have lived to regret it. Gray died at the hands of the posse, the other man, John Dowd, by suicide.

Shorty Gray, Tom Murry, Charles Taylor, the man with the gun. His aliases were endless, and at his death, when they had flung his body upon the floor of the prison morgue, they knew nothing of him. They had the name of a woman who had vanished, they had his clothes, the shreds of his mortal garment, not the man. Not even his place of birth. Nor had the posse dignified itself by thrusting his bloody head out of the train window to show the triumph of the law.

I used a borrowed car and drove in the direction of Fremont where my parents had lived when I was three. Near the little town of Gretna, close by, the posse had come upon the convicts. They had been free less than a week. My father had been right, they had no chance at all. The governor promptly proclaimed that the tragedy was the result of coddling convicts. The warden was buried with the eulogies of the church. The prison chaplain, by then ensconced in Tecumseh, refused comment. He had a new post. Perhaps he was wise.

So this was the end then, this leaf-filled autumn road where all that violence had consumed itself some sixty years ago. It had been winter and the last time Tom Murry had looked upon the day there had been only naked wind and the relentlessly advancing posse with its long-range rifles. That had been the last thing he had seen, the last, at least, the world knew about. But in a strange sense he had not ended there. A child, a solitary child from his own stark background, had remembered across sixty years.

There was no trace now of the blood and the anguish. It was dissolved in the snows of over sixty winters, the expended shells of the posse scattered in wood-rat middens, lost among corn shocks and the black seeds of the sunflowers. And I who carried the memory was also dissolving; the trade rats were exchanging me piece by piece. There was a pair of old wired rimless spectacles staring somewhere from the piled debris of a rat's nest that also contained the flint knife of an Indian. Deep in the midden was hidden a pine cone from a tree that no longer grew in the crevice where the glacial ice had lingered.

A thin aroma arose invisibly from the air—the time that the rats were collecting—a ball from a Sharp's buffalo gun, my own dust-grimed stare, we were all present, the essence of autumn, of carbon-14, disintegrated bits of measurable time at which the rat generations toiled incessantly while their middens grew. They were the finest collectors in the world, I thought, remembering that midnight desert struggle to regain the padlock.

Time. What did it matter now? It was breaking apart like disjointed memories. Soon I would be gone like Tom Murry and there would be nothing, nothing at all but a shimmering reflection in summer heat, or the midnight squeal of a rat carrying a stolen unusable watch, whose hands had stopped long ago, into a midden perpetuated by the efforts of scurrying generations. I grinned a little bleakly, thinking of it. Tom Murry grinned with me. We were all there in the mirage. Collectors' items. The time traders had a watch face, my glasses, and perhaps his gun.

I made another trip to the archives, mostly to say goodbye, goodbye to my awkward, stumbling childhood and the words my father had spoken. I gave the microfilm a final twirl till it sleeted past once more as Murry must have seen the world in his final moment, print, horsemen, and snow blurring away into an immensity that no longer contained shape or form.

"Thanks for the help," I said indifferently to the archivist. "You can put the *Herald* away now." I walked blindly to the door and vanished. The place would never see me again. The archivist would never know my name. No one would know that the blizzard had been resurrected, nor that snow had never ceased blowing through my mind since 1912. On the sign-out slip for the microfilm I had written Loren Eiseley, but on another crumpled in my pocket I had written Tom Murry. It was an unexplainable act of defiance. I would have turned it in, but a hand almost reached out and touched me. "The other alias is better," a whisper said. "We have to get out of this town." It was a voice from the past. I listened. I listened carefully; the past is very close to me.

CHAPTER 25

The Other Player

I AM coming to the end—the end of which I spoke in the beginning—the shattered mirror which can never be repaired but which lies in bits in the hallways of the mind itself. Feet crunch upon the glass as in an abandoned house; sometimes a ray of light strikes through a closed shutter and something still glitters, devastatingly beautiful, upon the floor. Or, similarly, a moonlit dream turns the fragments to soft shadows out of which come voices. I have not slept well since, in the hatchery days, I set the clock to wake me every hour. Perhaps I still fear the fire that might have consumed me there, or the dice game in the house of childhood. Who knows about these things? Who knows, sometimes in age, what one really is or if someone else—or alternately others—gazes from the eyes that we imagine are our own? Even psychologists admit to the reality of multiple personality.

But why philosophize? This is the story of a dream, but of a dream that came as I fumbled late amongst resurrected news clippings and objects that had been better left, like one's glasses and padlocks, to the hurrying time traders. In the words of the physicist Max Born, "the brain is a consummate piece of combinatorial mathematics." An hour comes when reality, the reality we know, gives way to combinations no longer causal or successive in character.

I have not dreamed such a dream before but something tells me I will dream again, though in a different fashion. It began, I think, in the slight fever of a cold. It went on, gaining in reality, until I could sense the wheels of a train jouncing beneath me and the lurch of the freight car in whose door I was standing. A harsh wind cut through the doorway but there was still a faint light. Strung out along the grade were men who seemed known to me from past decades. They were clothed mostly in the black of the earlier century. They ran and reached up gesticulating hands as though in warning or goodbye. The train was picking up speed for the storm-filled country ahead. I leaned against the door and watched with no regret the running figures fade. Fields and bridges were flashing by at an incredible pace. The whistle ahead alternately wailed and blasted over the desolate countryside. It reverberated mournfully in my head as I turned to compose myself for sleep on the hard floor of the car.

There were two of us in that boxcar and, looking on with astonishment, alternately inhabiting the mind of first one and then the other, I saw that one was a grey-haired man, the other a youth of eighteen. There was something familiar about both of them.

"There's always another town," ventured the eighteen-year-old as the cold cut deep into the car. "Find some packing paper and wrap it inside your shirt. Christ, man, try to sleep." He tightened his belt, turned, and heaved vainly at the drafty door which stuck and refused to close.

"Will there be other towns?" asked the older man a little simply.

"There will be a town," repeated another voice in the car. Something about it was suddenly vast and inescapably ominous. The Player, the Other Player, I thought wildly. The Player who had tried to burn me in that place of the life machines so long ago. The Player who had waited in the adjoining corridor and who never spoke. He was speaking now. "There will be a town," he said. The car filled with his presence. The train roared across a trestle that echoed like something heard in my youth. The Player no longer spoke in the car, but I knew he was there. "Jump," I whispered to the frightened eighteen-year-old whose hands shook with fear. "The train is doing sixty but you just might be the one to make it—the only one. Jump and make a fire for dawn."

We stood at the door in the black night and the beating rain, waiting for slackened speed and a soft embankment. Once more the youth hitched his belt. "Now," I shouted. He jumped but I never saw him hit the grade. The rain poured down in sheets.

The grey-haired man awoke in his bed staring at the ceiling in the dawn. The Player was still present but he had retreated to a corner as on that night in the hatchery.

"The young man jumped," I said after a long moment. "At least he had guts."

"You will not see him again," said the Other Player.

"And Feather?" I questioned, fingering names like beads.

"She does not remember."

"And the black-haired one in grade school whose name I cannot—?"

"No. It is too late—bad luck there, she will not remember."

"Did she make out?"

"Do you think it wise to ask?"

My head rolled back upon the pillow.

"Can you not be content?" said the Player. "You are an old man, a scholar now. You have come a long way through time. You have written books. You can name men since history began. You have stood in the places of the dead, handled their skulls. Content yourself. Anywhere along the way it could have been different."

"Or better," I said.

"But the women," I protested. "There was so little time," I

said. "I make faces, faces in pain when no one is looking. There were those who went away, or things which were not done right."

"You remember the game in the ruined house at evening? When you were a child? The score you could not read for sure?"

"Yes, yes I do."

"You played against all your possible futures. You played in the dim room at sunset. Remember?"

"I remember."

"You lost the unborn, remember?"

"I remember."

"You lost fame."

"I remember."

"And fortune."

"I remember."

"You left stones in various graves. The magic did not work. You are called a scientist?"

"I fought for them when all else was gone or laughed at. My kind have done it since the ice age. I would do it again."

"You were precocious at eighteen. You knew then how it would turn out."

"Yes," I murmured. For a moment I saw a line of squat, helmeted bodies and a goal ten yards behind me toward which the players surged. As the play began I could see the whole line shifting behind me.

"You were very fast," the voice conceded, "but your eyes were going even then. You had to turn to scholarship, remember? You liked this crush of bodies better. The direct approach. No metaphysics. At heart you are a primitive. As I remember, the game was lost."

"It was a game," I said.

I tried to face him down but the effort was like the breathing at the end of the field when the eyes of the fresh replacements are close and you know already how things will go.

"Would you like to play again for another ending?"

(261)

"Yes," I said. "No," I countered, considering. "I do not trust you. It might be worse. The pain—"

The Player waited patiently.

"You wish to play?" he asked again.

I nodded weakly. "It is the only way to win. The odds—"

The room grew suddenly quiet.

"Let us begin," I pleaded.

"There is only one game," said the remorseless gamester. "You have not learned its meaning. It is mentioned in your Bible. It is called the count of the days toward wisdom."

"And this is where it ends?" We were whispering now.

"Yes, this is where it ends."

I raised myself upon one elbow. "I will not play such a game twice," I said defiantly.

The Player paused, almost as if he shrugged invisibly.

"Others of your kind have found that best," he said finally. "They have taught themselves deliberately to leave the wheel of existence."

The clicking of the dice began again but it was far away.

"Who is playing?" I questioned.

"Hush," said the Player faintly, "I am going now. It is another game."

"Wait," I insisted. "I want to play."

"It is the only wisdom," brooded the Player. "That is the secret that was kept from you."

I was growing weaker now. "The play," I repeated. The clicking of the dice began again but the sound was faint. "I want to play."

"There is only the one game," affirmed the Player from somewhere behind me. "You play but once. That is why the days are counted. Lie back, you have already played."

"And it can be no different?" I asked. The Player did not answer. There was a great pain bursting in my heart. I needed the strength of the youth in the boxcar, but he was no longer there. He had—where was he now?

"He is gone," said the voice inexorably. "He never hit the grade, he is in you, wiped out in you."

I buried my face in the pillow. "Go away," I pleaded. "I do not want to remember."

"I am going," said the Player. "You have lost. There are new games beginning."

"But I won," I cried after him. "Remember, remember, that I played." It dawned suddenly morning in the harsh sunlight of Mexico, three decades back.

"That, too, is part of the wisdom," the voice came back to me. "You played. That is part of the counting. And this is where the kind of time that bewitched you began. Remember? *Behind nothing, before nothing.* This is the country of vertical time. I will leave you to add the zeros. The gods always carried them here."

My hands relaxed a little. I turned my head. I would sleep in the sun and consider. Carefully, I extracted a travel folder from my pocket and drew it over my face. It was lined with roads and towns crowded with other lives. I no longer believed the Player. Had I not been conditioned since childhood to escape? I would close my eyes and be patient as in that cabin long ago on the mountain in Colorado.

The clicking of the dice came a little nearer and then faded into a renewed dream. I was tired and I slept on the steps of a ruined temple. I cared no longer in what age I might awake. The Player was gone at last.

I turned and let the sun come faintly over my closed eyes. What had it all meant? I still did not know. Was this all that could be said of the counting? I tried to concentrate upon a face, but strangely that too was fading. I slept as the temple slept in the timeless Caribbean sun. This was what it meant then, the counting: the dots and bars on the great stelae. The wisdom could take care of itself. It was beyond me. It was beyond every man. But for all that the counting mattered.

THE OTHER PLAYER

Suddenly, inexplicably, I stood alone on a western hilltop in the falling snow of a blizzard that would never cease so long as the world remained. As though I were someone else I saw the approaching posse on horseback, their rifles black against the snow. Damn them, the snow had done it. I could go no farther. I dropped from the wagon. The lifted pistol in my hand did not waver. I was someone else now playing a role long finished because an innocent child could not forget.

I fired and the answering rifles brought me bloodily to my knees. I coughed. What mathematical equation was this? I had asked to play eternally against the Player. He had dissembled, cheated. It was forever a throw that turned out the same, eternally the same. A final mockery. Worship it the zero. But I would play, the warden and the hatred were long gone, but I would play. No man would beat me into line. I lifted the gun once more, the snow turning red beneath me. It was the last of the dream. *Behind nothing, before nothing, worship it the zero.* This, then, was the counting. It was the Player who controlled the dice.

I thought so in that dream. I do not believe it. "Between a man asleep and a man dead," Cervantes once wrote, "there is but little difference." Upon this point I would differ with the ironic assailant of windmills. I have said, and reiterated throughout this account of my journey, that either I have not slept at all, or I have lived a life of dreams so violent that at times I have struck out in defense, or striven with impotent muscles to beat back the powers of the dark. Curiously I have never dreamed of flight.

I have climbed up a solitary subway stair to a winking red glow upon a blank wall that denied me entrance, while a pushing crowd poured into a tunnel that I found utterly abhorrent and resolutely refused to enter. A sizable portion of my life has been given to such adventures which certainly do not equate with the conception of sleep as a little interim death. I have even started up unhesitatingly with the passage of a burglar's flashlight over my eyes, as though something couched and waiting had expected him there.

Always, since that headlong fall into unconsciousness long ago, I sleep high on two pillows, with my arms, even in the coldest weather, outside the covers. Do I still await the return of the laughing puppet? I do not know. I only know that I dream and the dream ends in that bloody violence in the snow, the time traders' exchange of sixty years ago. Why can they not effect another? I know that this is the screen I must penetrate, a screen to be whirled away into deeper snow as once I tried to do by spinning the microfilm viewer in the archives.

When my mother died, among her remaining sparse belongings was a satchel left to me. In it I had discovered a huge forgotten bone from my early diggings. The bone lies now upon my desk, massive and so mineralized it can be made to ring when struck. Why had she saved it? God alone knows, but I am aware there were once two of these ice-age bison forelimbs and that one lies far away in the burial of a dog named Wolf who wandered much with me and upon whom my head once rested by a fire.

I think we dreamed the same dreams, that dog and I. It was for this reason that I had seen, in the eyes of the man on the loading platform in Kansas, the great cold, the unutterable spaces, and the age from which he had come. We knew, rather than spoke, knew because the precise definitions of the present day did not exist between us, any more than they had existed between myself and the dog named Wolf now lying in his grave. We had not been dogs or people in any modern sense. We were merely creatures who hunted and shared together, products of a winter such as I had once glimpsed from afar across blinding icefields long ago.

I did not care for taxonomic definitions, that was the truth of it. I did not care to be a man, only a being. I lifted the huge bone meditatively. The dream had faded; there was no way to ask the Player to recast the dice. To do so always ended with Tom Murry, because Tom Murry had died as a hunted man yet still defined as human. I had inadvertently joined him, identified with him. But in that greater winter where I sought retreat, Tom Murry could lead me no further. I would continue to fall and die to no purpose. I remembered then the patient clicking that had followed me down the ancient vaults of that crumbling library through the door that out of long habit I had held open.

On impulse I put the bone beneath my pillow. I knew what I was doing. Wolf would help me, help me past that endless confrontation in the snow. The Player could not stop him, for we would be no longer man or dog, but creatures, creatures with no knowledge of contingency or games. All the carefully drawn human lines would be erased between us, the snows deeper, the posse floundering, the dice cup muted in the Player's hand. We would vanish together as an anonymous grey blur. The time traders would scurry to help us, even Coyote the trickster, who is unscrupulous and wins at gambling.

Make no mistake, I will dream again, but further, further back. The rifles will be silenced, the dice at last unshaken. I feel my hour coming. I am anxious to press on. They wait for me, the dog Wolf and the Indian, muffled in snow upon the altiplano.